Long-Term Care Investment Strategies

RONALD E. MILLS, EDITOR

Long-Term Care Investment Strategies

A Guide to Start-Ups, Facility Conversions and Strategic Alliances

A Healthcare 2000 Publication
IRWIN
Professional Publishing®
Chicago • London • Singapore

HEALTHCARE FINANCIAL MANAGEMENT ASSOCIATION

IRWIN Concerned about Our Environment

In recognition of the fact that our company is a large end-user of fragile yet replenishable resources, we at IRWIN can assure you that every effort is made to meet or exceed Environmental Protection Agency (EPA) recommendations and requirements for a "greener" workplace.

To preserve these natural assets, a number of environmental policies, both companywide and department-specific, have been implemented. From the use of 50% recycled paper in our textbooks to the printing of promotional materials with recycled stock and soy inks to our office paper recycling program, we are committed to reducing waste and replacing environmentally unsafe products with safer alternatives.

A Healthcare 2000 Publication

IRWIN Professional Publishing®

© Richard D. Irwin, a Times Mirror Higher Education Group, Inc., company, 1996

All rights reserved. No part of this publication may be reproduced, stored in a retrieval system, or transmitted, in any form or by any means, electronic, mechanical, photocopying, recording, or otherwise, without the prior written permission of the publisher.

This publication is designed to provide accurate and authoritative information in regard to the subject matter covered. It is sold with the understanding that neither the author nor the publisher is engaged in rendering legal, accounting, or other professional service. If legal advice or other expert assistance is required, the services of a competent professional person should be sought.

From a Declaration of Principles jointly adopted by a Committee of the American Bar Association and a Committee of Publishers.

Irwin Professional Book Team

Senior sponsoring editor: Kristine Rynne
Project editor: Christina Thornton-Villagomez
Production supervisor: Lara Feinberg
Senior desktop production supervisor: Andrea Rosenberg
Assistant manager, desktop services: Jon Christopher
Designer: DCD
Compositor: David Corona Design
Typeface: 11/13 Palatino
Printer: Braun-Brumfield, Inc.

**Times Mirror
Higher Education Group**

Library of Congress Cataloging-in-Publication Data

Long-term care investment strategies : a guide to start-ups, facility conversions and strategic alliances / Ronald E. Mills, editor.
   p. cm.
"A healthcare 2000 publication."
Includes bibliographical references and index.
ISBN 1-55738-622-6
   1. Long-term care facilities—United States—Finance—Planning.
   2. Long-term care facilities—United States—Planning. I. Mills, Ronald E.
RA997.L666   1996
362.1'6'0681—dc20   95-34923

Printed in the United States of America
1 2 3 4 5 6 7 8 9 0 BB 2 1 0 9 8 7 6 5

For Laura and Danny

Contents

Preface xiii
Acknowledgments xv

The Climate for Long-Term Care

1 *The Eldercare Market Today and Tomorrow*   1
Nancy Alfred Persily and Darolyn Jorgensen
   Who Are the Elderly and What Is Their Impact? 2
      Demographics 2
      Expenditures 3
      Social Resources 3
   What Is Long-Term Care? 4
      Health Status of the Elderly 4
      Economic Status 6
      Diversity among the Elderly 7
      Geographic Distribution 8
   The Continuum of Long-Term Care 9
      Temporary or Permanent Long-Term Care 9
      Community-Based Services 11
      Institutional Care 12
   Government Financing of Long-Term Care 17
      Medicare 17
      Diagnosis-Related Groups: A New Method of Reimbursement 18
      Resource-Based Relative Value Scale 19
      Medigap 19
      Medicaid 20
   Personal Financing Alternatives for Long-Term Care 21
      Medicare Managed Care 21
      Private Long-Term Care Insurance 22
      Health Individual Retirement Accounts 23
      Social HMOs 24
      Home Equity Conversions and Reverse Mortgages 24

Reaching Out to the Elder Market 25
Summary 26
References 28

2 Financing Alternatives for Long-Term Care and Senior Living Developments 31
James F. Sherman and Andrew P. Logan

Industry Demographics 32
Types of Senior Living and Long-Term Care Products 34
Development and Operating Statistics 36
The Financing Outlook for Senior Living and Long-Term Care 38
Today's Most Viable Financing Methods 40
   Equity Financing 40
   Debt Financing 43
   Credit Enhancements 46
Alternative Financing Methods 48
   Equity Financing 48
   Debt Financing 50
   Credit Enhancements 51
Summary 53
For Further Reading 53

3 Legal Perspectives on Long-Term Care Investing 55
Christopher J. Donovan

Joint Ventures 55
Venture Capital 56
Choice of Entity 57
Debt Financing 58
Regulatory Issues 59
   Licenses 59
   Letters and Reports 60
   Certificates 60

Long-Term Care Trends and the Law 61
   Integration 61
   Subacute Care 62
   Real Estate Mortgage Investment Conduits 62
Summary 63

4 *Technology and the Home Care Revolution*   65
Kathleen R. Crampton and Stephen B. Kaufman

The Home Care Industry Today 66
   Home Care Prescribers 66
   Home Care Providers 66
   Home Care Payers 68
The Growing Home Care Market 69
   Favorable Demographics 70
   Government Cost-Containment Efforts 71
   Managed Care 71
   High Technology in the Home 72
   Patient and Family Preference 72
   Chronic Conditions 72
   Problems of the Hospital Setting 72
   The Continuum of Care 73
Impending Changes in the Home Care Industry 73
   Cost Reduction and Maximization of Resources 73
   National Contracting Trends 74
   Requirements for More and Better Data 74
   Comprehensive Services 74
   Consolidation 74
   Diagnostic-Related Programs 75
Opportunities for Improved Care in the Home 75
   Improving the Availability of Healthcare Information 75
   Ensuring Medication Compliance 76
   Providing Support Services 76
   Maximizing Resources 76
   Connecting with the Information Highway 76
   Integrating Independent Devices 77
High-Technology Home Care Solutions 77
   Personal Emergency Response Systems 77
   Monitoring of Vital Signs 78
   Single-System Monitoring 78

Video Monitoring 78
Medication Reminders 79
Infusion Therapy 79
Voice-Activated Computer Aids 79
Device Networks: Putting It All Together 79
Summary 80
References 80

Strategies and Techniques

5 Strategic Facility Planning for Long-Term Care 83
James C. Morell

Long-Term Care as a Clinical Model 84
    Long-Term Care's Place on the Continuum 85
    How Customers and Suppliers View Long-Term Care 86
    How Payers View Long-Term Care 86
Long-Term Care as a Physical Location 87
    On-Campus Development in Existing Facilities 87
    On-Campus Development Requiring New Facilities 89
    Off-Campus Development 90
Determining a Market for Long-Term Care 92
    Product-Line Regulatory Mandates 92
    Sheltered Services Concept 93
    Service Continuum through an Integrated Network 93
Defining Market Expectations 94
    Patient and Family Expectations 94
    Providers' Expectations 96
    Payer and Employer Expectations 98
Identifying Solutions for Long-Term Care 99
    Case Study 1: Clinical Model Development 99
    Case Study 2: Operational Model Development 100
    Case Study 3: Physical Model Development 101
Implementing the Selected Solution 102
    Identifying Decision Criteria 102
    To Make or to Buy? 104
    Implementing Long-Term Care for a Managed Care Marketplace 105

Monitoring the Solution's Success   107
   Configuring Your Decision Dashboard   107
   When to Hold and When to Adjust   108
   Viable Alternatives   109
The Future of Long-Term Care   110
   Key Drivers for Long-Term Care   111
   Positioning Your Organization for the Future   111
   How to Build in Flexibility Today   112
Closing Observations   113
Reference   114

6 *The Appraisal of Long-Term Care Facilities*   115
Michael A. Lucas and Janet R. O'Toole

Establishing the Date of Value   115
Three Valuation Methods   116
   Cost Approach   117
   Sales Comparison Approach   120
   Income Approach   125
Assignment and Inspection   128
Design Features of Long-Term Care Facilities   131
The Appraiser's Output   133
Appraisal Standards and Types   133
   Appraisal Types   135
   Reporting Options   136
Appraiser Qualifications   137
State Regulation and Funding   139
Summary   140

7 *Information Solutions for the Long-Term Care Provider*   141
Marietta Donne Hance

The Need for New Information Systems Solutions   142
   The Reimbursement Landscape   142
   Integrated Long-Term Care   143
   Driving Forces in Long-Term Care   144
   Information Investments for the Future   149

Evaluation and Implementation of Automated
   Solutions   152
   The Information Vision   152
   Analyzing Operations   155
   Weighing Options   156
   Implementation   162
Keys to Success   164
References   170
For Further Reading   171
Acknowledgment   172

8 Demonstrating Financial Feasibility   173
Roy W. Byers

Purpose of the Financial Feasibility Study   174
Prospective Financial Information   174
Feasibility Study Reports   175
Feasibility Study Consultant's Services   177
Feasibility Study Process   179
   Step 1: Selection of the Feasibility Consultant   179
   *Step 2: Information Gathering and Preliminary
      Discussions*   180
   Step 3: Market Analysis   180
   Step 4: Financial Analysis   186
   Step 5: Examination Procedures   188
   Step 6: Report Preparation   189
   *Step 7: Review of Offering Statement and Other
      Legal Documents*   196
Summary   199

Resources   201
Investment Opportunities and News   201
Strategic and Financial Planning   202

Author Biographies   205
About the Editor   205
About the Authors   205

Index   211

Preface

Half primer, half manifesto, this book aspires to make plain sense of contemporary long-term healthcare and to arm prospective investors and developers with the fresh insights, encouragement, and practical information they need to seize the day.

Imagine a superior team of consultants assembled to design and implement a long-term care program of unprecedented vision and scope. Here they are—executives and managers, strategists and accountants, physicians and attorneys, innovators and implementers—covering the whole territory, from the sweeping arc of the planning horizon to that constricted neck of the funnel, the financial feasibility study. Our common purpose is to map the prevailing winds of change; to examine and, whenever possible, quantify the chief components of the long-term care investment; and to share proven strategies and techniques that launch ideas into action and actions into enterprises.

The concept for this book was straightforward enough: assemble top professionals doing important work in their respective disciplines, then let them tell what they know best. We begin with a survey of the long-term care landscape and end with practical, stepwise instructions for the astute planner and developer.

The first part of the book juxtaposes the distinct perspectives of the eldercare market, the investment community, the legal profession, and the new home care frontier. The second part introduces powerful strategies and techniques for planning facility construction or conversion projects, estimating the value of existing or proposed properties, choosing and implementing essential information systems, and demonstrating financial feasibility. Together, these eight chapters cover the basic whys and hows of a sound long-term care investment strategy.

Most publications of this sort rely on one beneficent firm for sole sponsorship and a suspiciously entangled roster of authors. Not so with this book. Call ours a mixed chorus. As editor, my obligation has been to select these clarion voices, to motivate them to write what they know, to preserve the fidelity of their arguments, and to cast every sentence in nearly standard American English. I have taken pains to give the authors plenty of elbow room to articulate their own outlooks, definitions, and solutions. Save these brief remarks in preface, I will let the contributors speak for themselves.

Ronald E. Mills

Acknowledgments

How my dear authors, in our bustling marketplace, found a moment's peace to reflect and compose their thoughts so lucidly is a marvel of character and concentration. My thanks to all of them.

As this yearlong project took shape, several reliable sources gladly referred me to willing authors and noted authorities. Special thanks to Linda Conheady and all at American Hospital Publishing, Inc.; to Rob Watson of the American Health Care Association; and to Bob Kramer, executive director, National Investment Conference for the Senior Living and Long Term Care Industries.

So many others along the way have shared their valuable time, insights, and kind words of encouragement. My heartfelt thanks to Christine Tabaka for the germ of an idea and, fondly, to Cathy Hankins for brave reportage from the battlefront. For my practical education in health care I am indeed grateful for the enlightening fellowship of my former colleagues at Chi Systems, Ernst & Whinney, and Ernst & Young; my current officemates at Ovation Healthcare Research; and my wonderful editors and publishers during the past dozen years.

And not a word would have been possible without the sustaining grace of my loyal clients. Most notably I am indebted to John Abendshien, for his mentoring and for giving me my flying start; to Dennis Patterson, for his generosity of spirit and trust; and to my friend Jeff Trotter, for the use of the hall.

Book manuscripts invariably rush to press trailing mild regrets and discarded pieces. Many individuals who early agreed to contribute chapters to this book either were unable to do so or contributed in other ways. I would be remiss not to mention, with gratitude, the contributions of Jo-Anne Andre, Jill Krueger, Marty Leinwand, Dr. Christine Mlot, Dr. Chuck Musfeldt, Brian Schiff, and Marilyn Woodrow. I especially offer thanks to Ken Kaufman, for my introduction to this fine publishing house.

Finally, my capable editors, Kris Rynne and Kevin Thornton, once again kept my tether from wrapping around the anchoring post, and the book is all the better for their gentle care and handling.

CHAPTER 1

The Eldercare Market Today and Tomorrow

Nancy Alfred Persily and Darolyn Jorgensen

The Graying of America, the Aging of America, our Aging Society, the Senior Market, the New Old Market, Eldercare, Senior Care, the Senior Political Bloc, Retirees, the Old, the Old Old, the Oldest Old, the Aging Woman, the Golden Years, the Silver Years, the Declining Years—however you label this phenomenon, the United States is experiencing a demographic imperative. With this imperative comes the need for long-term care services.

Social Security has had a great impact on older persons over the last 60 years. The orientation of minimum subsistence for the elderly changed to a replacement of income upon retirement from work at age 65.[1] This change made the older person, for the first time in our history, economically enfranchised. And in 1965 Congress enacted Medicare Parts A and B, an age-categorical medical insurance program providing those over 65 the opportunity to be medically enfranchised as well. This law, coupled with Title XIX, most commonly known as Medicaid, provided assurances that the children of older persons in need of long-term institutional care would no longer be burdened with that economic catastrophe, thus extending to the entire family the opportunity to be economically and medically enfranchised.

The concerns of the elderly have drifted in and out of the spotlight these last few decades. Today, however, senior citizens constitute a rapidly growing sector of our population, with more economic power than ever before in our history. Older people, on the average, use healthcare services more than any other age group. As they age they have increasing needs for helping services, such as homemaker care, transportation, shopping, and meal preparation, in order to remain independent. Such services are often paid for out-of-pocket.

This chapter examines the demographics of our aging population, describes the growing continuum of long-term caring services, and summarizes today's financing sources and alternatives, both public and private, for acute and long-term care. This discussion sets the stage and, most importantly, provides the reader with new information about the elderly and the continuum of care—information that goes way beyond the acute care model.

As Peter Drucker has said, "Management has no choice but to anticipate the future." In healthcare, we know that our future is the demographic imperative described in these pages. Now we must learn how to respond to this imperative in the most cost-effective and appropriate way to guarantee quality of life and continuity of care for all older Americans.

Who Are the Elderly and What Is Their Impact?

In recent years, long-term care has become an important issue for the public as well as for federal and state governments. To understand the future climate and market opportunities for long-term care, one must understand the three major driving forces—(1) demographics, (2) increases in governmental and personal expenditures, and (3) strains on social resources—that are putting long-term care in the spotlight today.

Demographics

The demographics of the elderly population are changing dramatically. In 1992, more than 32.3 million of our citizens were over the age of 65, and the U.S. Bureau of the Census estimates this number will increase to 70.2 million, approximately 23 percent of the population, by the year 2030. In addition, the population referred to as the oldest old, age 85 and over, will grow from its present three million to approximately nine million by 2030.[2]

The oldest old category has more women than men. Elderly women outnumber men by 3 to 2, and this disparity increases with age. For women 65 to 69 years of age in 1990, there were 81 men per 100 women; this number decreases by almost half, to 41 men per 100 women for those 85 to 89 years of age.[3] As a result, elderly women are more likely to end up widowed and living alone, and so they tend to suffer a disproportionately high level of poverty. Older men, on the other hand, often marry younger women or remarry after their spouses die and are thus more likely to have a caregiver during their later years.

Expenditures

Long-term care expenditures have represented the fastest-growing portion of expenditures in the healthcare sector. For example, between 1960 and 1991, nursing home expenditures increased almost 60 times, from $1 billion to $59.9 billion.[4] The growing number of elderly, particularly the oldest old, will cause an increased use of healthcare services. Since aging is often linked with disabling conditions, the current trend of increasing expenditures for long-term care is likely to continue.

Social Resources

The strain on social resources associated with caregiving can have grave implications for individuals and families. Caring for an older person clearly is not only a personal commitment but an economic one as well. Lost time at work equates to lost dollars for a company as well as for the employee caregiver. The Administration on Aging states that as many as one in four employees provides some form of care to an aged person.[5] The increasing numbers of women in the workplace, coupled with increasing single-parent families, make the stress on women, who are the dominant caregivers, even greater.

Increases in longevity often correlate to increased functional disabilities. Frequently, skilled caregivers are required to perform personal caregiving functions, such as dressing and bathing, and other tasks for older, disabled persons. These demands have major impacts on what caregivers can reasonably provide with or without outside assistance. As a result, caregivers today and in the future will seek long-term caring services along the continuum of care to improve the quality of life for themselves as well as for their elderly loved ones.

What Is Long-Term Care?

Long-term care can be defined as a set of health, personal care, and social services delivered over a sustained period of time to persons who have lost or never acquired some degree of functional capacity.[6]

To understand the settings and types of long-term services one receives, we must understand how the need for long-term care is measured. The service or care one receives is often measured by assessing limitations in a person's capacity to manage certain functions or activities. For example, chronic disability suffered after a stroke may result in the need for basic self-care, such as bathing, eating, dressing, going to the toilet, or getting in and out of bed. These kinds of activities are referred to as limitations in activities of daily living (ADLs). Limitations that reflect lower levels of disability are used to describe difficulties in performing functions such as meal preparation, grocery shopping, or managing money. These limitations are known as instrumental activities of daily living (IADLs).

Health Status of the Elderly

Despite the common belief that older people have poorer health, nearly 71 percent of elderly people living in the community describe their health as excellent, very good, or good, compared with others their age; only 29 percent report that their health is fair or poor.[7] Persons 65 and over are less likely to smoke, be overweight, drink, or report that stress has adversely affected their health than are younger persons. However, elderly individuals are far less likely to exercise regularly. Furthermore, their health perceptions are positively correlated with income. It is interesting to note, for instance, that approximately 26 percent of older people with incomes greater than $35,000, contrasted with only 10 percent of older people with low incomes (less than $10,000), rate their health as excellent.[8]

Although most elderly persons perceive themselves as healthy, health and the ability to function inevitably decline with advancing age. As individuals grow older, acute conditions become less frequent and chronic conditions become more prevalent. More than four of every five people aged 65 and older have at least one chronic condition, and the frequency of multiple chronic conditions increases with age, especially among older women. The most common chronic conditions among the elderly are arthritis, hypertension, hearing impairment, and heart disease.[9]

Heart disease is the leading cause of death and the leading diagnosis for short-term hospital stays and physician visits among people aged 65 and older. Despite the decline in death rates from heart disease, this disease still remains a severe health problem. Heart disease, cancer, and stroke together account for more than 75 percent of all deaths among the elderly.[10]

The degree to which an elderly person is affected by chronic conditions and diseases varies widely from person to person. For instance, one older person may be bedridden with arthritis, whereas another older person with a similar diagnosis may experience little discomfort and limited loss of mobility. One measure of the severity of such health conditions is the prevalence of limitations in ADLs and IADLs. These measures also are important indicators to determine the quality of life and the need for long-term care.

One of the drawbacks of using just ADLs as a measurement of health status is the lack of direct assessment of cognitive impairment. Any ADL deficits can be reflective of cognitive functioning but should not be used as the only determinant of cognitive impairment.[11]

The most widely recognized form of cognitive impairment or dementia is Alzheimer's disease. Alzheimer's disease is a slowly degenerative brain disorder that impairs memory and judgment and eventually results in degenerative physical conditions as well. The risk of this disease increases with age. Because the 85 and older population is the fastest growing segment of the population, it is evident why medical research and healthcare services recently have directed attention to this disease and those affected. Based on statistics from the Alzheimer's Disease and Related Disorders Association, an active group that has done an outstanding job of educating the public about Alzheimer's disease and other dementias, approximately four million Americans have this progressively debilitating disease; the number is expected to rise to eight million by the year 2000. Approximately 10 percent of the population 65 years of age and older are estimated to have Alzheimer's disease, and this percentage rises to 47.2 among those over the age of 85.[12] Because the diagnosis of Alzheimer's disease is difficult to make, even though many elderly persons have typical symptoms of the disease, a formal diagnosis most often cannot be made until an autopsy is performed. Thus, many believe it is probable that more older people suffer from Alzheimer's disease than any surveys indicate.

Economic Status

Seventy percent of the nation's personal wealth is controlled by the 50-plus population, with its $7 trillion in assets. The 10-year cohort with the highest percentage of discretionary income are those 65 to 74 years old, followed by the 55- to 64-year-olds. Although the dollars of this often-referenced $800 billion market are not as available for marketplace transactions as marketers would like to believe, the "young old," comprising the well elderly and those approaching retirement, have more discretionary income than any other age group. Currently, older people tend to be good savers because, many hypothesize, their generation was affected by the Great Depression. Much of this group's income goes into passive investments and not directly into the marketplace. However, because older people usually own homes that are paid for and are less likely to spend money on clothes and other luxury purchases, such as compact disk players, video cassette recorders, and televisions, they are more apt to spend money on services that make their lives easier, such as housekeeping, yard work, or a hairdresser.

Undoubtedly, the elderly have made great economic gains over the past decades. Much of these gains are attributable to a general increase in Social Security benefits between 1969 and 1972. Whereas prices increased by only 27 percent between 1968 and 1971, Social Security cost-of-living increases amounted to 43 percent.[13] The income gap between the elderly and nonelderly has narrowed ever since then. Even though the overall income of the nonelderly is higher, the elderly have the highest discretionary income of any age group and are less likely to be poor than the rest of the adult population.[14] Many analysts also assert that, when noncash resources such as government in-kind transfers, lifetime accumulations of wealth, and favorable tax treatment of income are considered, the average older person has overall economic resources roughly equivalent to those of the average younger adult.[15]

In the next decade, the cohort of elderly persons aged 85 and over will continue to experience rapid growth, but the number will remain relatively small in comparison to the younger elderly (ages 55 to 74).

For the next three decades, the group most likely to be well off, those 55 to 75 years old, will constitute approximately three-fourths of the mature market. There are, for example, more than 23 million persons aged 55 to 64 enjoying the highest per capita income of all 10-year cohorts; this number is expected to grow to 39 million by 2015, a 78 percent increase.[16] In 1988 approximately 63.6 percent of persons between ages 55 and 64 had a net worth of $50,000 or higher; over 43.1 percent of this same group had a net worth of $100,000 or greater. Although these proportions declined with age, more than half the elderly aged 75 and over still had a net worth of $50,000 or greater during the same period.[17]

Despite the overall improved economic picture for the elderly, it is imperative to realize that stark economic contrasts remain among subgroups of the elderly. There are, for example, significant differences between men and women and among different types of households. Average income statistics illuminate the polarization of these subgroups, with high-income elderly on one end and, on the other, a large cluster of older adults just above the poverty level, safeguarded only by Social Security. People aged 85 and older have significantly lower cash incomes than those who are aged 65 to 74, or 75 to 84. For example, the median cash family income of individuals aged 85-plus was less than three-fourths that of persons aged 65 to 74. In the South, more than one in five elderly persons is poor.[18] The oldest elderly are the most likely to have incomes below or just above the poverty level, and older women of every age group and every marital status are substantially more likely to be poor.[19]

Diversity among the Elderly

In the years ahead, there will be significant increases in the racial and ethnic diversity within the elderly population. Presently the nonwhite and Hispanic populations have a smaller proportion of aged than does the white population, but this phenomenon will change course beginning in the early part of the next century. Between 1990 and 2030, the older white population will grow by 92 percent, compared with 247 percent for the older black population and people of other races and 395 percent for older Hispanics.[20] As part of this trend, the number of older minorities who live alone also is projected to grow rapidly in the future. The number of black elderly (the second largest group of aged) living alone is expected to triple to 2.4 million individuals.

Consequently, as we begin the new century, our more heterogeneous elderly population will require bilingual and multicultural attention. This diversity will challenge policymakers, planners, health and social service providers, and marketers to address potential divisiveness within the elder community based on race and ethnicity, while also seeking ways to respond to the needs of all elderly persons.

Geographic Distribution

In 1992 approximately half (52 percent) of persons aged 65 and older lived in eight states: California, New York, Florida, Texas, Illinois, Ohio, Michigan, and New Jersey.[21] Also, in eight states—Alaska, Arizona, Florida, Hawaii, Nevada, New Mexico, North Carolina, and Utah—increases in the 85-plus population surpassed 50 percent.[22]

The apparent graying of the population in some states does not necessarily mean that elderly people are suddenly moving there. In fact, the majority of today's elderly population age in place, that is, they tend to stay where they spent most of their adult lives. Factors that have contributed to the high percentages of the elderly in some states are "countermigration" and the out-of-state migration of young adults.

Countermigration is a phenomenon that occurs when older people move to another state at retirement, then later move back home or to a state where family members live. Often countermigration occurs when a spouse dies or when the older person becomes sick or disabled. Findings from the Retirement Migration Project show that Florida, for example, lost significant numbers of elderly, primarily those more than 75 years old, to states outside the Sun Belt (Michigan, New York, Ohio, and Pennsylvania—all states that also send "young old" migrants to Florida).[23] In fact, Florida had the highest outward migration rate of any state for the population 75 years and older.

The outward migration of young adults is another important factor that has significantly contributed to the graying of some states. Young adults, in their early 20s, have the highest rate of residential mobility. Between March 1986 and March 1987, only 5 percent of older people moved, in contrast to 35 percent of 20- to 24-year-olds.[24]

The salient point for providers is that not all older people in the Snowbelt think in terms of moving to the Sun Belt. Many age in place or would prefer a long-term care setting in their hometown. A study further substantiating this phenomenon was conducted by Beverly Enterprises, a large nursing home chain. Nearly 19,000 family members and friends of residents in some 650 nursing homes owned by Beverly Enterprises were surveyed, and 46 percent of respondents indicated that the facility was chosen because it was nearby.[25] This tendency to remain close to family and friends applies both to the choice of a long-term care facility and to the preference to remain in a family home as long as possible.

The Continuum of Long-Term Care

Older persons in need of long-term care services, whether immediate or gradual, in the community or through institutions, require continuity of care—continued timely intervention with appropriate levels of services. Frequently the need for long-term care is triggered by a hospitalization, many times because of an acute flare-up of a chronic disease. Such a crisis is often accompanied by the mobilization of institutional and community resources with the goal of restoring patient functioning, personal health, and independence.

The need for case management and the assurance of continuity of care to foster movement out of the hospital in a responsible manner is paramount. Although speedy discharge is an economic imperative for hospitals, continuity of care has become a social and health imperative for persons in need of long-term care.

Because the elderly are becoming the opportunity market of the future, long-term care facilities and services must design new and innovative ways to meet the variety of needs, expectations, and health conditions of this population. An array of comprehensive services to address the health, social, and mental health needs of aged persons at all levels of intensity must be offered on both a short-term and a long-term basis.

Temporary or Permanent Long-Term Care

Stanley J. Brody divides the current market for long-term care services into two different groups: "those who require temporary support or short-term long-term care (STLTC), and those in need of permanent or extended support or long-term long-term care (LTLTC)." STLTC and LTLTC are differentiated by the level and duration of care needed and by the discharge status.[26] The findings of a recent study on health services utilization patterns indicate that approximately 45 percent of older nursing home residents stay in the nursing home less than three months.[27] This trend provides evidence of an existing large consumer group that uses STLTC services or "step-down" services, such as inpatient rehabilitation, convalescent care, restorative nursing care, or subacute care as well as community-based care, such as day care, home healthcare services, and outpatient rehabilitation.

Increased attention on rehabilitative services for the elderly is a relatively new phenomenon. Physicians and others are discovering that most older persons profit considerably from rehabilitation care. With decreasing lengths of hospital stay, more older people are being referred to inpatient rehabilitation units and nursing homes for short rehabilitative stays, especially after major joint replacements and strokes. Many hospitals and nursing facilities see opportunities to increase their market shares by offering rehabilitation services, which are reimbursable under Medicare and often supported by commercial insurers as well. These insurers recognize the cost-effectiveness such services provide by affording the elderly the opportunity to restore functioning or to keep them at a level of functioning that will help ensure greater mobility and independence.

Most recently, the delivery of subacute services has been in the forefront of long-term care. Subacute care focuses on the needs of patients who are not sick enough to stay in an acute care hospital but who require a higher level of skilled care than is traditionally delivered in nursing facilities. In essence, subacute care is a hybrid between acute and nursing home care. In fact, subacute units are being developed in both hospital and nursing home settings competing for this increasing market.

The top three subacute care services are rehabilitation, ventilator therapy, and intravenous therapy. According to the American Healthcare Association, several state Medicaid programs have established a special payment rate for subacute care provided in special units in nursing facilities. This open window of opportunity to increase nursing home profit margins is especially dear in an industry so dependent on Medicaid reimbursement.

Community-Based Services

Most older people are determined to live in their own communities as long as possible. A variety of services, including day care, home health and homemaker care, and case management, help older people accomplish that objective.

Adult day care. Adult day care is a community-based comprehensive program that provides a range of health and social services in a group setting during the day (less than 24 hours a day). Over the last two decades, the number of adult day care centers has grown dramatically, from fewer than a dozen centers that were supported by federal research and demonstration funds in the 1970s to an estimated 3,000 centers today.[28] Day care programs have basically two models: medical day care and social day care. Medical day care, as its name suggests, provides more medically oriented services, such as rehabilitation, special care (for example, for Alzheimer's disease), and nursing services. Social day care, on the other hand, provides older people with socialization in a safe environment, monitoring, meals, and social and mental stimulation. Most important, day care provides adjunctive care, which often affords the older person the opportunity to remain at home and/or to be cared for by a loved one for a longer time.

There is conflicting evidence about the benefits of adult day care programs for clients. Some studies have shown improvement in mental and physical functioning of adult day care clients, whereas others have concluded little effect. With the growth of the elderly population and the stresses associated with caregiving, continued growth in the adult day care market appears inevitable.

Home healthcare. Home healthcare recently has been recognized as a major industry in the long-term care marketplace. According to the National Association for Home Care, in 1992 there were approximately 11,300 organizations across the country providing home-delivered healthcare services. Home healthcare includes all professional and paraprofessional long-term care services that are provided in the home—either the recipient's or a caregiver's home. Home care embraces physician services at one end of the continuum of cost and skill, and housekeeping or meal preparation at the other end.[29] The increased utilization of home care can be attributed to the aging of the population; the shift from cost-based reimbursement to a prospective payment system in hospitals, thus encouraging early discharge; the expansion of health maintenance organizations (HMOs); and technological advances affording patients the opportunity to have services such as respiratory or intravenous therapy at home rather than in a hospital or nursing home setting.

As noted by Shawn Bloom of the American Association of Homes for the Aging, "There are a lot of home health agencies across the U.S., but few are part of a larger continuum." This statement suggests that home healthcare is and should be a part of a continuum of care requiring case management, the management of resources, assessment of the home and family, and constant monitoring of services.

Home healthcare is the fastest-growing healthcare sector and gives the elderly yet another alternative to institutional care. It also represents an opportunity market to serve those who prefer to age in place and remain independent in the community for as long as possible.

Institutional Care

The list of alternative institutional settings for the elderly has grown by leaps and bounds in recent years. Beyond traditional nursing homes, the elderly can now receive long-term care services in special care units, assisted living facilities, continuing care retirement centers, or Alzheimer's disease units.

Nursing homes. Nursing homes provide nursing care, medical services, and personal care services to persons unable to care for themselves due to illness or infirmity. Typical services include nursing care (both restorative and palliative care), preventive and health maintenance services, rehabilitation, and ancillary and supportive social services.[30]

Although nursing home care is the service most closely associated with long-term care, only about 5 percent of people aged 65 and older are in nursing homes at any one time.[31] However, it is estimated that 43 percent of people who were 65 years old in 1990 will use nursing homes at some point in their lives.[32] According to data from the National Health Provider Inventory (NHPI), in 1991 there were 1,615,686 nursing home beds in 15,511 U.S. facilities, 90 percent of which had between 25 and 199 beds.[33] The NHPI findings are summarized in Table 1-1.

TABLE 1-1 Number of Facilities, Beds, and Residents in U.S. Nursing Homes and Board-and-Care Homes, 1991

Selected Characteristic	Nursing Homes*			Board-and-Care Homes†		
	Facilities	Beds	Residents	Facilities	Beds	Residents
Total	14,744	1,559,394	1,426,320	31,431	482,650	413,040
Bed size						
Less than 25 beds	563	8,101	7,105	27,377	204,816	179,473
25–99 beds	7,041	456,704	418,225	3,352	160,883	138,035
100–199 beds	6,028	792,348	724,966	544	71,863	59,605
200 beds or more	1,112	302,241	276,024	158	45,088	35,927
Ownership						
Profit	10,522	1,086,907	984,560	19,726	309,469	259,041
Nonprofit	3,497	372,272	348,090	9,694	143,142	128,516
Government	725	100,215	93,670	2,011	30,039	25,483
Geographic region						
Northeast	2,654	328,435	312,864	5,660	110,359	97,134
Midwest	5,137	518,917	468,636	8,817	105,515	91,742
South	4,708	503,522	457,944	7,090	131,982	111,840
West	2,245	208,520	186,876	9,864	134,794	112,324

*Excludes hospital-based facilities.
†Excludes 8,578 nonresponding board-and-care homes.

The National Health Provider Inventory reveals several interesting trends, especially concerning the living arrangements of the elderly, that marketers and planners should consider before constructing new facilities. For example, the Midwest and the South had twice as many nursing homes and almost twice as many beds as the Northeast and the West combined. Consequently, the Midwestern states have the highest proportion of elderly living in nursing homes, a small proportion in board-and-care homes, and even fewer relying on home health services.[35] In the West, the elderly use board-and-care homes more than any other region, and they use nursing homes and home healthcare less than any other region. Finally, elderly in the Northeast rely more on home healthcare than do their contemporaries in other regions, and they also rely heavily on board-and-care and nursing homes. Although the need for nursing home care was similar in each region, the various regional differences could be due to geography, a lack of nursing home beds, the higher cost of nursing home care, the availability of alternatives such as board-and-care homes, and, most important, the Medicaid policies in various states.

Medicaid reimbursement and eligibility varies considerably from state to state, as do other services covered under state Medicaid programs. In some states, Medicaid covers the costs of approved assisted living programs, provides payments to family members who wish to be full- or part-time caregivers in their own homes, and provides funding for foster care for older persons, among other programs. These creative funding initiatives cut down the use of institutional care in many states.

Nevertheless, the future demand and market for nursing homes appears solid. With the growing number of people aged 85-plus having three or more ADL deficits and requiring skilled care for longer periods, coupled with increased prevalence of Alzheimer's disease, it is very likely that the nursing home population will continue to increase, although perhaps not at the same speed it has, due to both alternative programming and the increasingly better health of older persons.

With other levels of care now being fueled through the increasing market for home health services and assisted living facilities, tomorrow's nursing home residents will enter the facility sicker, with more functional disabilities. In addition, more emphasis has been placed on innovative facility design features to create a more homelike atmosphere and a better quality of life. For example, Horace D'Angelo, Jr., president of Caretel Corporation, created the first "care-eh-tel," which combines the best of the healthcare and hospitality industries. He threw out conventional designs featuring long corridors, hospital room design, and overall institutional appearance. Residents have restaurant-style menu choices and eat in a dining room designed as a place to entertain visitors as well.[35]

The bottom line for marketers and developers is that the future eldercare market will prefer settings that offer choice, and without the institutional image and ambience. One must question, however, if such amenities truly will be evaluated by future residents or if, in fact, caregivers are the ones who must be impressed when selecting facilities for their loved ones. Since older people entering nursing homes today are sicker, many with dementia, most often they do not evaluate the facilities themselves.

Special care units (SCUs). SCUs, the Alzheimer's Association has realized, are unregulated in most states, and few states have reporting requirements. The Alzheimer's Association rightly questions the degree to which such units are truly special. The issue was recently brought to the attention of public and private sectors by the American Association of Retired Persons. As a result, there have been recent efforts on behalf of the states to enact legislation and implement guidelines for nursing homes that have designated units for residents with Alzheimer's disease and other dementias.

Of equal consideration are the criteria for admission to Alzheimer's units. Many nursing homes report that as many as 80 percent of their residents have diagnoses of Alzheimer's or other forms of dementia. What nursing homes often need are special care units for residents who are physically disabled but not cognitively impaired. Studies conducted by Persily and Blume indicate that these residents' quality of life and ADLs increase when they are clustered in special care units with residents who have similar disabilities but are not cognitively impaired.

Assisted living. Assisted living, a new market-driven housing alternative for the elderly and disabled persons, can be defined as any residential group program that (1) is not licensed as a nursing home, (2) provides personal care to persons with the need for some assistance in daily living, and (3) can respond to unscheduled needs for assistance.[36] A national study by the University of Minnesota's Term-Care Decisions Resource Center determined that assisted living has the potential to save 20 to 40 percent over the cost of nursing home care, providing at the same time increased autonomy and enhanced quality of life for residents.[37]

A key issue for assisted living facilities is where to place them on the continuum of care. Currently, there is no agreement as to whether such facilities should be identified as providing a distinct level of care a step below nursing homes or whether a substantial overlap with nursing homes should be expected. This is a major concern for developers. If assisted living facilities are viewed as a nursing home substitute, federal and state regulations could affect the return on investment.

An important finding of the Minnesota study was that, in general, there was an apparent inconsistency between facility developers, who sought a market niche below nursing home level, and the group of residents they actually attracted. According to administrators in the study, residents admitted were frailer than developers had anticipated. Thus, developers were serving a target population more disabled than originally expected. Undoubtedly, market research plays a key role in determining the success or failure of any long-term care setting. Developers absolutely must perform adequate market research before initiating the design of facilities and services.

Since 1989, Oregon has gone further than any other state in appropriately developing assisted living facilities to meet the needs of its elderly population. Oregon's facilities have promoted independence, privacy, and at the same time accommodated residents with disabilities comparable to those of many residents in nursing homes.[38] Growth of assisted living facilities has been spurred by regulations, incentives to developers, and a reimbursement source for the care component of the program. In reviewing Oregon's experience, it becomes clear that, by applying marketing research methods, establishing clear goals, and working closely with key actors, such as lenders, state architects, and health departments, an assisted living facility can indeed prove to be a highly successful venture.

Continuing care retirement communities (CCRCs). CCRCs combine independent living in a campuslike setting with access to nursing home care. Unlike nursing homes or assisted living facilities, CCRCs cater to recent retirees who at the time of entry usually are independent, with minimal functional disabilities.

A unique characteristic of CCRCs is that, when an elderly person needs higher levels of care, the CCRC is able to transfer the resident to a setting within the community that offers skilled care, such as a nursing home. With such lifetime access to healthcare services, CCRCs resemble long-term care insurance policies that are financed through a combination of an entry fee and a monthly maintenance fee. However, the high costs of CCRCs often make them accessible to only the affluent elderly. The important point is that CCRCs offer another alternative along the continuum of care available to the elderly who wish to ensure access to healthcare and support services in their later years.

Key marketing strategies of CCRCs and other similar long-term care settings should communicate to the consumer how the services and programs offered enhance quality of life. Despite the heavy focus on add-on services (meals, housekeeping, etc.), such services per se are not the strong selling points commonly believed, as David Wolfe points out in his book *Marketing to Baby Boomers and Beyond*.[39] Marketing surveys that focus on the graying market have indicated that older consumers are more interested in how products increase life satisfaction than they are in the luxury apartment or services themselves.

Alzheimer's disease units. Alzheimer's disease units emerged during the 1980s as a care option for people with Alzheimer's disease and related dementias. A national survey indicates that approximately one in 10 nursing homes has a special unit or program for people with dementia.[40] Often nursing homes charge more for care in Alzheimer's units, and many have found them to be a major marketing asset for their homes. The growing 85-plus elderly population, who are most likely to develop Alzheimer's disease and other forms of dementia, are a prime target market for these units.

Government Financing of Long-Term Care

Our federal and state governments finance selected long-term care services through the Medicare and Medicaid programs. Eligibility requirements, payment mechanisms, and specific program components have changed considerably over the years.

Medicare

An entitlement program enacted in 1965 to provide health insurance for persons 65 years of age and over and for certain other disabled individuals, Medicare consists of two parts: the Hospital Insurance program (Part A) and the Supplemental Insurance program (Part B).

Medicare Part A provides coverage for inpatient hospital services, up to 100 days of posthospital skilled nursing facility (SNF) care, home health services, and hospice care. Patients must pay a deductible ($696 in 1994) each time their hospital admission begins a benefit period. Medicare pays the remaining costs for the first 60 days of care. Beneficiaries requiring care beyond 60 days are subject to additional charges. Patients requiring SNF care are subject to a daily coinsurance charge for the 21st through the 100th day ($87 in 1994). There are no cost-sharing charges for home healthcare, and there are limited charges for hospice care.

Medicare Part B is voluntary and provides coverage for physician services, laboratory services, durable medical equipment, and outpatient hospital services. Any persons enrolled in Part A may enroll in Part B simply by paying a monthly premium ($41.10 in 1994).

It is important to note that Medicare does not provide reimbursement for all care needed by many elderly individuals. Older persons often do not understand what benefits are covered and frequently assume that they are well insured for long-term care as well as acute care services. True, outpatient prescription drug costs and physician charges above the Medicare allowable rate are often covered in whole or in part by Medigap insurance, but out-of-pocket costs are still prohibitively high for many older persons. An attempt was made, with the passage of the Medicare Catastrophic Care Act of 1988, to cover the high out-of-pocket expenses that many Medicare beneficiaries encounter. The law was repealed, however, when many elderly persons realized that they had to contribute additional monies to be insured for the additional coverage.

Diagnosis-Related Groups: A New Method of Reimbursement

During the early 1980s medical costs continued to rise faster than the rate of inflation. In response, Congress enacted the prospective payment system (PPS) in October 1983. Under PPS, fixed hospital payments are established in advance of the provision of services on the basis of a patient's diagnosis. In effect, the Health Care Financing Administration (HCFA), the administrative agency of the Department of Health and Human Services, changed from a passive payer to an active controller of hospital costs.

The system's fixed prices are determined in advance on a cost-per-case basis, using a classification system of 487 diagnosis-related groups (DRGs). Each Medicare case is assigned to one of the 487 DRGs based on the patient's medical condition and treatment. Psychiatric, long-term care, and rehabilitation beds are excluded from DRGs.

Because the DRG system provides a fixed payment for each episode of illness, the incentives for the hospital are quite different than under the traditional retrospective payment system. Under the DRG system, hospitals have strong incentives to decrease the number of days in the hospital, perform fewer unnecessary services, and reduce inefficiencies in order to increase income. Although there are no data to substantiate a diminishing quality of care since the implementation of DRGs, hospitals have been accused of "cost shifting" and discharging patients "quicker and sicker," which has led to alleged "patient dumping" into nursing facilities. Indeed, the shortage of nursing home beds was created almost instantly when DRGs were implemented, but more recently the growing number of facilities and improved home health services and discharge planning have led to an increased availability of nursing home beds.

Resource-Based Relative Value Scale

Medicare also has recently developed a PPS, known as the Resource-Based Relative Value Scale (RBRVS), especially for physicians. This scale compares the relative work involved in performing one service with the work involved in providing other physicians' services. The relative value is the sum of three components: physician's work, practice expense, and malpractice expense. Each physician service is assigned its own relative value.

In the past, Medicare reimbursed physicians on the basis of reasonable and customary charges. Now, under the RBRVS system, each physician is reimbursed according to a set fee schedule. A physician may choose whether to accept assignment on a claim. In the case of an assigned claim, Medicare pays the physician 80 percent of the approved amount. The participating physician can bill the beneficiary only the 20 percent coinsurance plus any unmet deductible. Physicians who do not agree to accept assignment on all Medicare claims in a given year receive only 95 percent of the recognized amount paid to participating physicians.

Nonparticipating physicians may charge beneficiaries more than the fee schedule amount on nonassigned claims, and these balance billing charges are subject to certain limits. Considerable attention has been paid to nonparticipating physicians who bill patients in excess of the limiting charges. The publicity has been very negative for physicians, causing not a few to change their behavior and accept assignment.

Medigap

Medigap is private health insurance designed to supplement Medicare coverage. Approximately 65 percent of persons enrolled in

Medicare today have some type of supplemental insurance.[41] Over recent years, increasing costs of healthcare coupled with large deductibles have raised the cost of Medigap insurance, leading some companies to drop such coverage for their retirees and causing many older people to let their supplemental health insurance coverage lapse just when their needs increase. Private insurers have been hesitant to offer policies that do more than provide protection against the copayments for the limited services Medicare covers. Consequently, many elderly persons have found it particularly difficult or entirely unaffordable to acquire policies that cover just a limited number of nursing home days, home healthcare benefits, prescription drugs, and physician costs beyond the Medicare-approved rates.

Medicaid

Enacted in 1965 and initiated in 1966 as a joint federal-state program that pays for medical services for low-income persons, Medicaid is administered by the states with partial federal funding. Each state establishes its own eligibility, coverage, and reimbursement standards within broad federal guidelines. Thus, there is considerable variation among the states in types of services covered and the amount of care provided under specific service categories.

Because Medicare covers only short-term skilled nursing home care following a hospitalization, beneficiaries who need long-term care and do not have long-term care insurance face catastrophic costs. As a result, many elderly individuals are forced to seek assistance through their state Medicaid programs once they have depleted their personal assets. In fiscal year 1993, long-term care expenditures accounted for 30.3 percent of total Medicaid expenditures.

Medicaid eligibility requirements vary by state. Medicaid is a means test-triggered entitlement, meaning that one must first "spend down" his or her assets—deplete virtually all accumulated personal resources—to qualify for Medicaid benefits. This means test is not difficult for most older and disabled persons to pass, given that the average cost of nursing home care is approximately $45,000 per year. In fact, several state studies have assessed the percentage of Medicaid recipients living in nursing homes who were not eligible for Medicaid when they were originally admitted, revealing that between 27 and 45 percent of residents "spent down" while in nursing homes.

In 1993, as reported by the Department of Health and Human Services, the nation spent $59 billion for nursing home care. The Medicaid program and out-of-pocket payments account for nearly 90 percent of this total, reflecting the severely limited private insurance coverage for long-term care benefits today.

Government financing of long-term care is an issue many expect to become more pressing, given current demographic trends and projections of utilization of long-term care services by the elderly, particularly the oldest old.

Personal Financing Alternatives for Long-Term Care

Despite much effort, the failure of the federal government to implement any type of health care reform, including long-term care coverage, puts pressure on individual states and on the private sector to take action. The goals of new and innovative financing alternatives are to provide protection against financial risk and to extend benefit levels beyond those offered in Medicare. Currently, some financing options that address such goals are Medicare managed care, long-term care insurance, individual retirement accounts, social health maintenance organizations, and home equity conversions and reverse mortgages.

Medicare Managed Care

Medicare managed care is becoming one of the hottest new products for HMOs. Through a Medicare managed care plan, HMOs offer the elderly subscriber the opportunity to receive all of the coverage Medicare provides plus extras, such as prescription drugs, eyeglasses, and even discounts on dental care. Best of all, there are no deductibles, coinsurance, or premiums for a Medigap policy for the consumer.

The HCFA has licensed HMOs to develop programs for Medicare patients. For each person who joins, the government pays the HMO 95 percent of the average cost of care for a Medicare patient of similar bands (age and sex) in the same region. For organizations experienced in managing care for their enrollees, this program has been a financial bonanza. The growth of Medicare managed care has been most dramatic in the West, where managed care growth in general is most rapid. Florida has been another major growth area for Medicare HMOs.

In 1987 there were three million Medicare enrollees in HMOs, and enrollment is projected to exceed 6.6 million by the end of 1995. Even though HMOs are required to accept all "comers," enrollees tend to be the healthiest in their bands. The government is studying ways to adjust payments and to offer older people different models such as Medicare Select. A 14-state experiment, Medicare Select allows beneficiaries to buy Medigap insurance at discounted rates if they enter a managed care plan. HCFA also is investigating a preferred provider organization (PPO) option to slowly move older enrollees into managed care. In short, some form of managed care for most Medicare enrollees is more a reality now than ever before.

Private Long-Term Care Insurance

Long-term care insurance is relatively new and costly, so only about 5 percent of older Americans have purchased this type of coverage. However, the market is growing rapidly. According to the Health Insurance Association of America (HIAA), the number of long-term care insurance policies sold increased from 815,000 in 1987 to 2,430,000 in 1991.[42] Studies indicate that the private insurance marketplace will find the growing number of relatively healthy older persons to be an attractive market and a reasonable insurance risk pool. Still, the broad-based affordability of long-term care insurance has not been demonstrated for the entire population.

With the growth of long-term care insurance, insurers and policymakers have raised many complicating issues. Insurers are concerned about the potential for adverse selection, whereby only those who need care actually buy insurance, and about the moral hazard problem, whereby individuals decide to use more services than they otherwise would because they have insurance. On the other hand, policymakers are concerned about the quality of such policies and the affordability of long-term care insurance.

Individual states therefore have begun exploring the potential role of public-private insurance partnerships in financing long-term care. The goal of such a partnership is to make policies more affordable, to provide individuals who need long-term care an alternative to out-of-pocket spending and Medicaid spend-down, and to protect personal assets.

The Robert Wood Johnson Foundation has awarded planning grants to develop ways in which partnerships should be structured. Between August 1991 and April 1993, the HCFA approved Medicaid plan amendments from five states—California, Connecticut, Indiana, Iowa, and New York—which, in effect, create a partnership between Medicaid and private long-term care insurance in those states.[43] In general, the plan amendments allow the states to establish more liberal assets standards for persons with private insurance coverage. In other words, each dollar that the insurance policy has paid out is subtracted from the assets the individual still owns and is not required to be spent down. For example, a resident of Connecticut who purchases a long-term care insurance policy that covers and pays out $50,000 of nursing home benefits will be able to retain those assets and gain Medicaid eligibility for coverage of any additional nursing home care that he or she needs.

The ultimate impact of this public-private approach on the marketability of private insurance is unclear, since operating experience at the present time is limited. Nevertheless, action taken on behalf of the state of Connecticut illustrates existing opportunities and ways that individual states can develop plans to assist in the reform of long-term care financing.

Health Individual Retirement Accounts

The establishment of individual retirement accounts (IRAs) is yet another alternative for individuals to finance and plan for eldercare. Individuals can contribute up to $2,000 annually to IRA accounts in banks, mutual funds, or insurance companies. The interest is not taxable until the individual withdraws the funds between the ages of 59½ and 70. Earlier withdrawal is subject to major penalties on the interest.

Tax reform legislation of 1986 restricted the deduction of IRA contributions. For persons whose employers also offer retirement savings programs, only those individuals whose adjusted gross income is less than $35,000 (for couples, less than $50,000) can deduct any IRA contributions from their income tax.

Social HMOs

The social health maintenance organization seeks to voluntarily enroll persons 65 years of age and over in an innovative prepaid program that integrates the delivery of acute and long-term care services.[44] Central to the social HMO concept is the merger of the managed care and HMO models through capitation financing and provider risk sharing.

The problem of prepaid long-term care services is that providers are at some financial risk, and in the case of long-term care there will always be lingering uncertainties about lengths of stay, scope of services to be used, and adequacy of funding. So far, social HMOs are merely HCFA demonstration programs. One of the largest issues has been member recruitment; some elderly persons prefer to retain their own physicians and, presently, the concept of a social HMO seems difficult to explain to the elderly market. We currently lack sufficient data to accurately analyze social HMO efficacy or efficiency. If the HCFA demonstrations prove successful, the social HMO could well become an important alternative to the financing of acute and long-term care for the elderly.

Home Equity Conversions and Reverse Mortgages

The most commonly held and valuable assets of the elderly are their homes. Three of every four persons over the age of 65 own their own homes, and recent statistics indicate that 80 percent of these do not have a mortgage.[45] However, many elderly individuals are "house-rich but cash-poor." In fact, the American Housing Survey in 1987 indicated that 19 percent of poor, elderly homeowners had homes valued at $70,000 or more.[46]

For this reason, home equity conversions and reverse mortgages, which permit aged homeowners to use existing home equity to pay for healthcare services, have proved to be a valuable option for financing long-term care. An equity conversion involves the sale of the home to a third-party investor, with stipulations that allow the person to remain in the home for a specific number of years into the future, or until death. Funds from the sale of the house are used to pay for needed long-term care services, so that elderly homeowners need not give up their home.

Reverse mortgages allow homeowners to borrow against home equity with no repayment of principal or interest due until the end of a specified term, until the home is sold, or until the borrower dies. This plan can provide a single lump-sum payout to the borrower or monthly payments for a given term.

Leo Baldwin, president of LEO, Inc., a consulting firm based in Washington, D.C., says that this type of financing "is a very substantial market, but it's got to be cultivated slowly. The idea of a person 'eating' into his home equity is not an easy thing to get used to at first, regardless of the financial benefits on a month-to-month basis."[47]

Reaching Out to the Elder Market

The shrinkage of the younger markets and the knowledge that some 70 percent of the nation's wealth lies with the 50-plus population have convinced businesses to take notice of older consumers and to develop new ways of addressing the graying of America.

Older persons today are more educated and more physically active, live longer, and are more active consumers than ever before. They are taking charge of their health-related issues rather than submitting to someone else's choices. To successfully market to the elderly, we must respond to their desires to "be all that they can be." Older consumers dislike products or services that emphasize their limitations or adversely reflect their self-images. They often think of themselves as looking and acting at least 10 years younger than their chronological age. A retirement housing complex, for example, marketed as a place for "old" people in ads showing older, gray-haired, inactive persons, will not fare well in the eldercare market. Retirement communities instead should be marketed and operated to emphasize opportunities for remaining active—mentally, physically, and socially.

Marketing messages aimed at the maturity market should not be things-oriented but rather experience-oriented. Because older adults are no longer as involved in a career and raising a family, material things do not hold as much personal value to them as do opportunities to promote and sustain inner growth.

Consumer behavior, like other demographic trends, is strikingly different in various subgroups of the elderly population. Younger people, aged 40 or under, are in a stage of life when the acquisition of possessions not only serves a purpose but defines who they are and provides evidence of what they have accomplished. People in the third stage of their lives, those aged 40 to 60 years, tend to spend money on catered experiences, such as restaurants, cruises, housecleaning services, sports events, or interior decorating services. For persons aged 60 to 80, there is yet another transition to a stage in which possessions and services provide less satisfaction and life experiences become more important. This age group most likely will spend its discretionary income on such things as educational magazines, theater and concert tickets, travel to a foreign country, or adult education classes.

In the final analysis, then, the successful marketer will study the older adult population not as a single group but as various subgroups of a rapidly growing population. Additionally, to effectively target those aged 60 to 80, marketers need to associate the products and services they promote with experiences that enhance life. Thus, long-term care settings and services that offer choice, promote continuity, stress independence, and emphasize positive experiences will prove to be successful in the eldercare market of today and tomorrow.

Summary

The senior market will only grow larger and more important in the years to come. The baby boomers—those fortunate enough to have no personal recollection of the Great Depression—are more likely to spend money, access resources, and become challenging consumers.

Older persons, however, do not stand alone; the family is an important part of the equation. Only recently have we seen families with four or in some cases even five generations of living relatives. Older people usually are members of families, and long-term care often involves family decision making. In addressing the eldercare market, we must consider the needs of family members young and old.

Proximity to children's homes is one of the major criteria for choosing a nursing home. Keeping in touch with children and grandchildren are major marketing tools used by airlines and travel companies. Long-term care is a family affair, too, and goods and services provided to older persons should often be promoted to caregivers as well.

Where do we go from here? We must recognize that the U.S. market for goods and services is no longer in the hands of the young. Products and services that focus on older persons and their families are a market for the future. We must now organize our management and marketing expertise to respond to the needs of those who will be our elder Americans of the next century. The time to make that leap is now.

References

1. Brody, S. J. "Strategic Planning: The Catastrophic Approach," *The Gerontologist,* 1986, vol. 26, no. 4, p. 3.
2. *A Profile of Older Americans: 1993.* Washington, D.C.: American Association of Retired Persons, 1993, p. 1.
3. *A Profile,* p. 1.
4. U.S. Dept. of Health and Human Services and National Center for Health Statistics. *Highlights of Trends in the Health of Older Americans: United States 1994.* Washington, D.C.: USDHHS, 1994.
5. *Elder Facts: Business and Aging.* Washington, D.C.: Administration on Aging, 1994.
6. Kane, R. A. and R. L. Kane. *Long-Term Care: Principles, Programs, and Policies.* New York: Springer Publishing Company, 1987, p. 5.
7. U.S. Senate Special Committee on Aging. *Aging America: Trends and Projections: 1991 Edition.* Washington, D.C.: U.S. Government Printing Office, 1991, p. 7.
8. Special Committee on Aging. *Aging America,* p. 8.
9. Special Committee on Aging. *Aging America,* p. 5.
10. Special Committee on Aging. *Aging America,* p. 2.
11. Persily, N. A. *Eldercare.* Chicago: American Hospital Publishing, 1991, p. 78.
12. *Alzheimer's Fact Sheet.* Chicago: Alzheimer's Disease and Related Disorders Association, 1991.
13. Special Committee on Aging. *Aging America,* p. 10.
14. Persily. *Eldercare,* p. 92.
15. Special Committee on Aging. *Aging America,* p. 10.
16. Wolfe, D. *Marketing to Baby Boomers and Beyond.* New York: McGraw-Hill, 1993, p. 121.
17. Wolfe. *Baby Boomers,* p. 130.
18. Special Committee on Aging. *Aging America,* p. 15.

19. Special Committee on Aging. *Aging America*, p. 16.
20. Special Committee on Aging. *Aging America*, p. 20.
21. Special Committee on Aging. *Aging America*, p. 21.
22. "Oldest Old Population Soars by 50% in Some States," *McKnight's Long-Term Care News* 15, no. 3 (March 1994), pp. 3, 20.
23. Longino, C. F., Jr.; J. C. Biggar; C. B. Flynn; and R. F. Wiseman. *The Retirement Migration Project*. Miami: Center for Social Research in Aging, University of Miami, September 1994, p. 5.
24. U.S. Bureau of the Census. "Geographical Mobility: March 1986 to March 1987." *Current Population Reports* Series P-20, No. 430, April 1989.
25. "Consumers Choose LTC Facilities Nearby: Beverly Study," *McKnight's Long-Term Care News* 15, no. 9 (September 1994), p. 2.
26. Brody, S. J. and J. S. Magel. "Step-Down Care Promotes Vertical Integration." *Hospitals*, December 1985, pp. 76–77.
27. Spence D. A. and J. M. Weiner. "Nursing Home Length of Stay Patterns: Results from the 1985 National Nursing Home Survey." *The Gerontologist* 30, no. 1 (1990), p. 16.
28. O'Shaughnessy, C. *Adult Daycare: A Fact Sheet*. Washington, D.C: Congressional Research Service, September 29, 1994.
29. O'Shaughnessy. *Adult Daycare*.
30. Persily. *Eldercare*, p. 160.
31. Special Committee on Aging. *Aging America*, p. 162.
32. Murtaugh, C.; P. Kemper; and B. Spillman. "The Risk of Nursing Home Use in Later Life." *Medical Care* 28, no. 10 (October 1990), p. 36.
33. Sirrocco, A. *Vital and Health Statistics*. Advance Data, Centers for Disease Control and Prevention/National Center for Health Statistics publication no. 244. Washington, D.C.: U.S. Government Printing Office, February 23, 1994.
34. Sirrocco. *Vital and Health Statistics*.
35. Wolfe. *Baby Boomers*, p. 104.
36. Kane, R. A. and K. W. Wilson. *Assisted Living in the United States: A New Paradigm for Residential Care for Frail Older Persons?* American Association of Retired Persons, 1993.
37. Kane and Wilson. *Assisted Living*.
38. Kane and Wilson. *Assisted Living*.
39. Wolfe. *Baby Boomers*, p. 173.
40. National Institute on Aging. *Progress Report on Alzheimer's Disease*. (Internal Document) Bethesda, Maryland: National Institutes of Health, 1994.
41. Special Committee on Aging. *Aging America*, p. 151.
42. *Highlights of HIAA Long-Term Care Market Survey*. Washington, D.C.: Health Insurance Association of America, 1992.

43. Price, R. *Medicaid: Long-Term Care and the Elderly.* Washington, D.C.: Congressional Research Service, August 25, 1993.
44. U.S. Senate Special Committee on Aging. *Developments in Aging: 1992.* Washington, D.C.: U.S. Government Printing Office, vol. 2, p. 45.
45. Special Committee on Aging. *Developments in Aging,* vol. 1, p. 285.
46. U.S. Bureau of the Census and U.S. Department of Housing and Urban Development. "American Housing Survey for the United States in 1987," *Current Housing Reports* H-150-87, December 1989.
47. Wolfe. *Baby Boomers,* p. 275.

CHAPTER 2

Financing Alternatives for Long-Term Care and Senior Living Developments

James F. Sherman and P. Andrew Logan

As the American population ages, so grows our need for specialized care facilities and housing for senior adults. Across the country, a wide variety of healthcare and residential facilities are being developed to meet the increasing consumer demand. Most recently, however, financing for senior living and long-term care developments has been difficult to find. Some projects have been credit enhanced or rated to facilitate the financing process, but credit enhancement and project rating are usually available only to larger, financially strong organizations.

Factors limiting the availability of financing in the conventional or public markets include the awareness of projects that have experienced or are experiencing financial and/or operational problems, the limited knowledge of senior adults' psychosocial characteristics and long-term care needs, and the increasing scrutiny of "real estate" loans by lending institutions. Although most banks continue to classify long-term care projects as real estate, in reality most such loans—especially loans for developing retirement centers and assisted living facilities—are made to larger organizations. Still, most lenders have avoided financing specialized senior health care and housing developments for the following reasons:

- Historically, the industry has been "tainted" by defaults and poor operations, especially life care communities, many of which were poorly planned. It should be noted, however, that the default rate and incidence of poor management among nursing homes, assisted living facilities, and nonprofit senior living communities has been relatively low.

- Operations of these facilities are complex, being a troublesome mixture of real estate, hospitality, and healthcare.
- The buildings are often specialty, single-purpose structures that can be difficult to convert to other uses.
- Lenders fear the negative publicity they might face if forced to foreclose on a senior living facility.

This chapter defines today's most viable methods of financing through equity, debt, and credit enhancements. It also contains a directory of alternative financing methods to assist developers in obtaining funds despite the investment community's general reluctance to back ventures in the senior living and long-term care industries.

Industry Demographics

Population aging is one of the most significant American demographic trends of this century. Already, 25 percent of our citizens are over the age of 50 years. Some of the other more notable demographic facts and figures include the following:

- Approximately 6,000 Americans turn 65 years old each day. By 1996 baby boomers will begin to turn 50 years old at the rate of one every 18 seconds (29,000 per year).
- The elderly population grew twice as fast as the rest of the population in the last two decades.
- The "very old" population (85 and over) is expected to grow sevenfold by the middle of the next century.
- The population aged 50 and older control 77 percent of the country's household financial holdings—stocks, bonds, and liquid assets.

As delineated in Figure 2–1, by the year 1990 Americans aged 65 and over numbered 31,799,000, constituting 12.7 percent of the U.S. population. In 1900 this age group numbered only 3,084,000, or 4 percent of the population. Of further significance is the rise in population of those aged 85 and over. In 1900 this group numbered 123,000, or 0.2 percent of the population, compared with the 1990 census of 3,461,000, or 1.4 percent of the population.

FIGURE 2–1 Aging U.S. Population: Population 55 Years and Over, 1900–2050

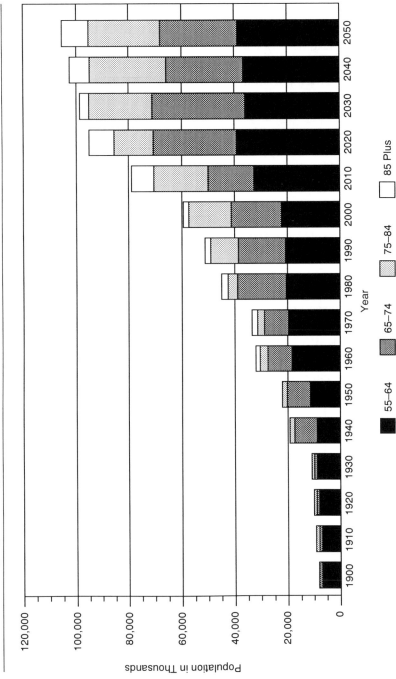

Source: U.S. Census of Population, 1890–1980, and projections of the population of the United States: 1983–2080. *Current Population Reports*, Series P-25, No. 952 middle series.

A recent U.S. General Accounting Office report indicates that the number of senior adults needing assistance with activities of daily living will double during the next 30 years. In 1965 approximately 6.5 million senior adults needed assistance with eating, cooking, dressing, hygiene, and other personal activities. By the year 2020, some 14.3 million senior adults are expected to need such assistance. The report also estimates that 40 percent of the senior population do not get the care they need because of a shortage of nursing home beds and alternative living arrangements and the lack of substantial government financing for home care.

Types of Senior Living and Long-Term Care Products

The long-term care industry lacks standardized definitions for the continuum of real estate, hospitality, and healthcare services marketed primarily to the senior adult. Nonprofit sponsors have been marketing a variety of senior living and long-term care products to older adults since the 1800s. With the implementation of the Medicaid and Medicare programs in the 1960s, many proprietary organizations entered the long-term care industry. Only within the last 10 years have proprietary developers begun developing senior living projects.

Table 2–1 lists the wide variety of contemporary senior living and long-term care developments.

The growing variety of senior living and long-term care developments has created fragmentation within the healthcare delivery system for senior adults, leading to confusion for consumers seeking appropriate living arrangements. When gauging the viability of future projects, developers must ask themselves several critical questions:

1. Is the financial arrangement acceptable and affordable?
2. Does the product design and mix meet the needs of prospective residents or patients?
3. For long-term care developments, does the current and future reimbursement from insurance and entitlement programs cover operations and capital costs? Furthermore, are there enough private-paying patients in the service area

TABLE 2-1 Types of Senior Living and Long-Term Care Products

Senior Living without Healthcare	Senior Living with Healthcare	Long-Term Care Developments
Rental apartments	Continuing care retirement communities (CCRCs)	Skilled and intermediate nursing homes
Adult congregate living facilities	Life care communities	Convalescent hospitals
Congregate living facilities	Any housing that offers healthcare with various financing arrangements, including:	Assisted living facilities
Resort retirement communities		Domiciliaries
Condominiums		Board and care homes
Co-operatives		Rest homes
	Refundable or nonrefundable lifetime lease with monthly service fees	Personal care facilities
		Residential care facilities
	Refundable or nonrefundable entrance fees with monthly service fees	
	Refundable or nonrefundable endowment fees with monthly service fees	
	Rental	
	Condominiums	
	Co-operatives	
	Stock purchase	

to cross-subsidize potential losses from other types of patients?

4. In senior living developments, is there a real and/or perceived need for a healthcare component?
5. How important to the senior population is the identity of the sponsor or developer?

Obviously, in-depth research will be necessary to determine the type of senior living and long-term care products that will be acceptable in your local community and to determine whether the development of such a project is financially feasible.

Development and Operating Statistics

Significant operating differences within the same product lines make it difficult to discuss development and operating statistics for long-term care products in general terms. The various factors that influence not only the type of product to be constructed but also the riskiness of development include the following:

- Sponsorship
- Product design and location
- Need-generated demand versus a discretionary decision
- Product pricing, including the availability of third-party payers
- Experience of the project's marketing, management, developer, and architecture team.
- Financing, especially the debt/equity ratio.

Table 2–2 provides statistical ranges that may be useful in assessing the development and/or operation of various senior living and long-term care products. These ranges should be used only as guidelines for analytical purposes and should not be relied on as absolutes. All such products and businesses will, of course, be influenced by certain local and regional factors, such as wage differentials, impact and zoning requirements, and operating cost differences attributable to climate, supply costs, and other factors.

Need-generated products appear to absorb their units faster than those requiring a discretionary decision by the consumer. However, senior living developments without a healthcare component are usually less management- and operations-intensive.

Various industry surveys indicate that the capital formation methods for senior living and long-term care products are beginning to change rapidly. In the past, most developments were financed primarily through conventional loans, tax-exempt bonds, and equity/limited partnerships. Due to changes in the tax and banking laws and in general economic conditions, we can expect to see the following:

- Higher equity requirements.
- Increased scrutiny of market research and financial analyses for proposed projects.

TABLE 2-2 Long-Term Care Industry Developmental and Operational Statistics

Statistic	Senior Living Developments without Healthcare	Senior Living Developments with Healthcare	Long-Term Care Developments — Assisted Living Facilities	Long-Term Care Developments — Nursing Homes
Average development costs*	$65,000–$75,000 per unit	$75,000–$85,000 per unit	$65,000–$85,000 per unit†	$25,000–$30,000 per bed
Average monthly absorption rate‡	1.5–2.0	2.0–3.0	3.0–4.0	7.0–8.0
Average stabilized occupancy	90%	93%	92%	95%
% of expenses to gross revenue	45%–50%	50%–55%	55%–70%	75%–80%
Average payer class revenues§				
Private pay/insurance	100%	100%	95%	25%
Medicare				6%
Medicaid			5%	64%
Managed care/Veterans Administration				5%
Average monthly rate per person				
Semiprivate			$1,217	
Studio (private)	$900–$1,000	$900–$1,000‖	$1,947	
1 bedroom	$1,000–$1,100	$1,000–$1,100‖		
2 bedroom	$1,200–$1,400	$1,200–$1,400‖		
Skilled care				$2,677
Unskilled care				$2,373
% attraction of residents or patients by service area#				
Primary	50%	50%	65%	80%
Secondary	25%	25%	25%	15%
Tertiary	25%	25%	10%	5%

*These figures do not include all "soft" costs such as marketing, operating deficits, etc.
†Most of the recently constructed assisted living facilities are residential by design and therefore are categorized by per unit versus per bed cost.
‡Absorption is defined as the number of units filled at stabilized occupancy divided by the number of months from initial marketing. This rate does not account for turnover during this period. All private-pay nursing homes normally have a much lower absorption rate. Absorption rates vary greatly from project to project.
§Only private-pay developments were considered, with the exception of nursing homes.
‖Utilization of the healthcare component not included. Most communities have a separate charge structure for this component.
#Service areas by location and product type can vary greatly
Sources: Marion Merrell Dow, *Managed Care Digest/Long-Term Care Edition*, 1993; American Association of Retired Persons, *Changing Needs for Long-Term Care: A Chartbook*; "An Overview of the Assisted Living Industry," *Assisted Living Facilities of America*, October 1993.

- Requirements for experienced marketing, management, and development teams.
- Expectations that sponsors or owners be financially committed to the development.
- Higher capitalization rates and debt service coverage in analyzing the value of a property.

Despite these trends, the forecast for investors in senior living and long-term care developments is only partly cloudy.

The Financing Outlook for Senior Living and Long-Term Care

Although there are a wide range of possibilities in financing senior living and long-term care projects, many conventional lenders will likely continue to avoid financing these specialty products. Luckily, there are many other sources of financing, as listed in Table 2–3.

Because many markets are overbuilt for congregate living arrangements, sponsors and developers today would do well to explore the market for assisted living arrangements, including stand-alone assisted living facilities. The U.S. Federal Housing Administration (FHA)'s Section 232 Board and Care Mortgage Insurance program provides credit enhancement for assisted living facilities. The intent of this program is to insure up to 90 percent of the real estate portion of a loan for a for-profit organization. Sponsors should therefore secure financing for the noninsured portion of the loan, required reserves, start-up marketing costs, and initial operating deficits. For some projects, these funds can be significant.

Expect to see aggressive pension funds and lenders willing to "participate" in the future cash flows of proposed projects and to provide discounted interest rates and closing costs in return for such participation. In addition to a 10 to 20 percent equity contribution (or bank letter of credit), developers should be fully prepared to present their lender with a rigorous financial feasibility study and market assessment. The comprehensive loan package should include the following:

- Sponsorship information, including resumes and financial statements.

TABLE 2-3 Funding Sources for Senior Adult Housing Developments

	Primary Sources	
Equity	**Debt**	**Credit Enhancements**
Limited partnerships	Tax-exempt bonds IRS Section 142(d)	FHA mortgage insurance HUD 232, Expanded
Sponsorship	Qualified 501(c)(3) bonds	Letters of credit
Joint ventures	Securitization of existing mortgages	
Equity arrangements	Seller financing	
	Real estate investment trusts (REITs)	
	Fannie Mae	

	Secondary Sources	
Developer financing	Taxable bonds	FHA insurance
Participating mortgage fund	Conventional lending	HUD Sections 221(d)(4) and 213
Venture capital	Foreign lending	Private mortgage insurance
Syndication		FSLIC insured bonds
Investor equity		Corporate surety/guarantee
		Guaranteed investment contracts (GICs)
		Put/option bonds

- Project concept overview.
- Market study, with a consumer survey in most cases.
- Member of the Appraisal Institute (MAI) appraisal and/or financial feasibility study.
- Site location and site studies.
- Building plans.
- Cost of project.
- Financial projections, including rent-up and cash flow projections.
- Management and marketing agreements.
- Details of loan being requested.

The bottom line is that viable sources of financing for the long-term care industry have been and will continue to be limited. Developers and sponsors who do not listen to the desires and needs of their local market will have difficultly filling and maintaining a census should they find financing. Today's successes or failures will ultimately set the stage for the types of financing available to the industry in the future.

Most conventional lenders will continue to require substantial equity from sponsors for development projects. Since most lenders treat long-term care and senior living developments as real estate loans, sponsors may consider separating the shelter and service areas to obtain more favorable financing terms. Alternatively, the service package could be sold to a financially stronger joint venture partner uninterested in the real estate component of the development. Either way, detailed legal arrangements to protect all parties are of paramount importance.

Although sources of financing today are limited, our aging population will continue to drive entrepreneurial development, ultimately creating even more ways of providing shelter and services for senior adults. The availability of more and better sources of capital in the future will surely depend on the integrity and business savvy of those sponsors entering the market today.

Today's Most Viable Financing Methods

We here summarize the industry's most promising financing methods, listed as primary funding sources in Table 2–3 and categorized as equity or debt financing vehicles or credit enhancements.

Equity Financing

Limited partnership. A frequently used form of legal ownership for long-term care communities and retirement facilities is the limited partnership. Limited partnerships are often the preferred form of ownership when passive equity investors' money is used or when a diverse group of partners or individual investors pool their resources and rely on the expertise of a developer as general partner. Three factors account for the popularity of the limited partnership: (1) limited partnerships, like general partnerships, are subject to a single level of income tax; (2) similar to corporate shareholders, limited partners have limited liability; and (3) the general partner can retain control of the project.

Sponsored joint ventures. In sponsored joint ventures, traditionally through nonprofit organizations, the sponsor forms either a proprietary or nonprofit subsidiary that develops, manages, and funds facility development. The parent sponsor provides cash infusions to the project, over which the specially formed subsidiary maintains control.

In joint ventures with local sponsors—as in the partnership of an experienced developer with a local entity, such as a church or hospital—both the sponsor and the partner contribute equity, while the developer provides the expertise necessary to develop the senior living facility. One caveat to such a partnership is the necessity of resolving nonmarket issues before development. Such nonmarket issues typically include a sponsor's desire to serve the community or to provide ministry versus the developer's profit motives. Fortunately, these two seemingly contradictory goals are often compatible when planned for accordingly.

For a developer willing to trade control for national experience and quick financing, a viable alterative is a joint venture with a national sponsor. National sponsorship is an option for proprietary as well as nonprofit organizations. National sponsors contribute tried and tested products (facility design, operations, and financing methods), national name recognition and exposure, and credit enhancements, which lower the cost of debt. As in a local sponsorship, however, it is paramount to have a clear understanding of each partner's goals and objectives before entering into such a relationship.

Equity arrangements. The traditional equity arrangements by which life care communities have raised equity require resident-paid, nonrefundable endowments or entrance fees. This up-front "contribution" (in addition to monthly rent and maintenance fees) covered all of the resident's housing and healthcare needs for life, with the exception of acute hospital care and physician services.

Actuarial errors and excessive healthcare cost increases caused the failure of a number of such communities and have given rise to four alternatives to the life care concept: (1) condominium ownership, (2) co-operatives, (3) membership organizations, and (4) lifetime lease fees. Realizing the potential financial exposure from contracting care for life, the trend among life-care communities has been to limit the total care included under resident contracts.

Essentially identical to ownership in conventional condominium housing, condominium ownership in a retirement community provides the developer with an immediate return of both capital and profit. The disadvantage to this form of ownership is the potential loss of operational control over the facility, taxation of profits, and the lack of control over second sales of the individual units. Sponsors' control can be maintained by limiting ownership rights through management contracts, deed restrictions, and certain other covenants. However, the goals of the condominium association membership must be considered carefully when addressing such issues.

In a co-operative, a corporation owns the retirement community and residents buy shares of the corporation (represented by stock certificates). The primary financing advantage of a co-operative format is the overall corporate ownership of the facility. Another advantage for developers is that some co-operative arrangements qualify for financing in conjunction with federal housing subsidy programs. A co-op offers lower closing costs, less regulation, and lower taxes than does condo ownership. The main disadvantage to developers brought by this form of legal ownership relates to issues with the co-operative board or governing body, similar to the problems faced by condominiums.

By granting residents the right to use a long-term care facility's services, the developer of a membership organization maintains control over the property, much like a country club does. The sponsor also maintains control of the membership and operations through rules and regulations adopted by the sponsor and agreed to by the members. Profit may be realized through the provision of services as well as by selling memberships to persons who meet the criteria for admission (based on age and/or economic status).

Providing residential accommodations on every level of the continuum of aging, in addition to meals and healthcare services, a continuing care retirement community (CCRC) typically offers its citizens the opportunity to reside in apartments, assisted living units, skilled nursing beds, and occasionally acute care beds in hospitals. Residents may pay large entrance fees and monthly maintenance fees and, in exchange (similar to life care), the CCRC agrees to provide care for a given number of years or at a lifetime discounted rate. Some portion of the entrance fee may also be refunded to the estate or survivors at death, depending on the contractual arrangement.

Debt Financing

Tax-exempt bonds. Affordable debt financing for assisted living and long-term care developers and operators has long been scarce. Having enjoyed greater access to tax-exempt bonds, nonprofit organizations have consistently had the advantage over their for-profit competitors. Section 142(d) of the Internal Revenue Service (IRS) Code levels the playing field, creating new opportunities for tax-exempt financing designed to encourage investors to develop facilities to match growth of the elderly population. For-profit entities that meet the traditional "low to moderate income" and "residential" criteria of the Housing and Urban Development Agency (HUD) can now access the tax-exempt bond markets to finance individual or pooled projects of $5 million or more. The process usually takes between 90 and 120 days to complete, with fees ranging from 4 percent to 6.5 percent. Credit enhancements are available through HUD 232 mortgage insurance or a bank letter of credit.

Resident income requirements for tax-exempt bond financing are as follows:

- 20 percent of the units must be reserved for residents with incomes at 50 percent of the area's median, or 40 percent of the units must be reserved for residents with incomes at 60 percent of the area's median.
- Family size lowers income amounts; a family of four meets the 60 percent criterion, while a single individual can make only 42 percent of the median income to qualify.
- Residents' incomes must be checked annually, and those making more than 140 percent of the area limit will no longer qualify.

Resident income requirements are discontinued after the bonds are retired or after 15 years, whichever comes first. The amount that a resident pays for client-support services is not limited to his or her earned income alone. "Spending down" of assets; sporadic gifts from family members; and income from tax-exempt bonds, inheritances, insurance payments, and capital gains in excess of the area median income limit may be used to pay monthly rent.

Regarding the residential criterion, to qualify for tax-exempt status, each unit in a funded facility must include living, sleeping, eating, cooking, and sanitation facilities.

A primary advantage to tax-exempt financing of a long-term care facility is the ability to finance almost all of the project's hard costs. Financing can be either fixed or variable rate, and there is flexibility in structuring the debt retirement schedule. The bond issues are long-term (maturing in 20 to 40 years) in a combination of serial and term bonds. Similar to balloon payments, the term bonds include an annual sinking fund for partial or full payment. Serial and term bonds normally are structured with a coupon rate that allows for a level debt service schedule.

Debt service payments on tax-exempt bonds can be initially capitalized from cash reserves, bond proceeds, or a letter of credit, and terms typically require a 1.30 or greater debt service coverage ratio.

Qualified 501(c)(3) bonds. Long-term care facilities and retirement communities that operate as qualified 501(c)(3) entities and that are exempt from federal taxes may be financed through tax-exempt bonds. If operated by benevolent organizations, such facilities must meet four criteria for tax-exempt bond status: (1) all property to be provided by the net proceeds of the bond issue must be owned by a 501(c)(3) organization or governmental unit; (2) not more than 5 percent of the security for the bonds may be used for private business purposes; (3) not more than 5 percent of the net proceeds may be expended on property used in a private business; and (4) the aggregate amount of qualified bond debt owed by the debtor 501(c)(3) organization (and any other 501(c)(3) organization under common management with the debtor) and outstanding debt must not exceed $150 million. These criteria are further defined in Section 145 of the Federal Tax Code.

Securitization of existing mortgages. The securitization of existing mortgages entails packaging a portfolio of performing mortgages and selling them on the secondary market as rated securities. The securitization market is relatively new to long-term care; however, it has become very active since 1992 and is available for borrowers of all sizes. Debt service coverage ratios are typically 1.4 for skilled nursing facilities and assisted living facilities, versus 1.3 for congregate living facilities.

The securitization proceeds as follows. A Wall Street investment/banking firm makes a mortgage to an operator. Those loans are aggregated with a pool of like loans (for long-term care, skilled nursing facilities, assisted living facilities, etc.) to attain critical mass. The cash flows are paid on the first of each month and then passed through to certificate "bond" holders. The pool is taken to a national rating agency and classified by risk and sized by class. Finally, the investment firm sells the pooled bonds to the appropriate buyer based on allocation of risk.

Securitization of existing mortgages can be classified by the allocation of risk as either principal transactions or best effort transactions. In a principal transaction, the borrower gets the funds immediately and the security firm bears the risk of the sale and spread. In a best effort transaction, the investment firm merely acts as a financial advisor and broker of the investment instrument. In both cases the borrower is responsible for the cost of the transaction.

Seller financing. Developers can save time and money through seller financing, when a seller is willing to accept a portion—usually 10 to 30 percent—of the purchase price in the form of a mortgage-backed note secured by the property, as either a primary or secondary mortgage or a wraparound. Developers can avoid lengthy and expensive property appraisals and loan committee approval when financing through a seller who is familiar with the property.

Seller financing holds benefits for both buyer and seller: (1) it may be the only way for a borrower with limited resources to purchase the land or property; (2) under the Internal Revenue Code's installment sales provision, the seller can avoid paying capital gains taxes in the year of the sale; and (3) the seller may realize higher returns on takeback financing than presently available in the market, in addition to receiving a fixed income over an extended period.

It is important that the buyer knows the reason behind a seller's willingness to finance the purchase, in order to structure the transaction to the greatest benefit of both parties. If the seller is financing for tax purposes, the buyer should thoroughly understand the Internal Revenue Code's installment sales provision. Conversely, if the seller is retiring and interested in fixed income, the buyer should compare the proposed financing rate against other long-term investments such as bonds, annuities, or certificates of deposits (CDs).

Real estate investment trusts (REITs). A REIT is a corporation, trust, or association (other than a real estate syndication) that invests in real estate through purchases or mortgage loans. REITs traditionally have allowed small investors to participate in large, quality real estate investments, which are professionally chosen and managed.

Although REITs can tie with venture capital funds as the most expensive form of capital, they do have three major advantages: (1) REITs typically provide 100 percent financing; (2) sale/lease-back transactions require interest-only payments, resulting in excess cash flows to owners; and (3) REITs will finance even those who operate only two or three facilities.

REITs are governed by the Securities Exchange Commission, the state in which they are organized or headquartered, the North American Securities Administration Association, and the stock exchange on which they are traded. REITs can be organized either as trusts managed by a board of directors or as corporations.

A REIT's external management, commonly known as an advisor, provides the REIT with certain management services. The advisor does the following: (1) researches, develops, and markets investment opportunities; (2) consults with the directors on buying, selling, or evaluating investment opportunities; (3) acts as the REIT's legal agent in real estate sales, purchases, foreclosures, and claim settlements; and (4) oversees the companies hired to manage REIT-owned properties.

Credit Enhancements

Two important credit enhancements, HUD Section 232 board-and-care mortgage insurance and letters of credit, can allow borrowers to use the credit ratings of others or to in other ways decrease their cost of borrowing.

FHA mortgage insurance. Before the Housing and Urban-Rural Recovery Act of 1983, Section 232, FHA mortgage insurance was limited to nursing homes and intermediate care facilities. The act amended the Section 232 program to permit federal mortgage insurance to assist in financing the construction or improvement of board-and-care homes (to include domiciliary care, adult congregate living, assisted living, personal care, and residential care homes). Section 232 now covers facilities that provide services for individuals who cannot live independently but who do not require the services of a skilled nursing facility—those who need, "room, board, and continuous protective oversight." On November 29, 1994, HUD expanded the Section 232 program to allow the purchase or refinancing of existing facilities originally financed either through the 232 program or through other vehicles. The revised mortgage insurance program covers existing homes, intermediate care facilities, and assisted living facilities and excludes the previous requirement for substantial rehabilitation.

Besides being less expensive and having more favorable terms than other mortgage insurance products, FHA mortgage insurance offers several other key advantages. First and foremost, the notes include low-interest, fixed-rate, fully assumable loans. The loans are nonrecourse with liberal prepayment options. To qualify for FHA guarantee, board-and-care homes may be privately owned and operated for profit or owned by a private nonprofit organization. Maximum mortgage amounts for proprietary organizations are limited by a loan-to-value ratio of 90 percent of hard assets, including the physical plant and major movable equipment.

Projects (and insurable mortgage amounts) are differentiated by project status as either new construction or rehabilitation. The maximum mortgage term for Section 232 is 40 years. The mortgagor is responsible for monthly principal, interest, and mortgage insurance as well as for all taxes, property insurance, and any special assessments or ground rents.

Further details regarding HUD Section 232 can be found in the *HUD Handbook,* Section 4600.1 and Notice 86-20 (HUD).

Letters of credit. Commercial banks issue letters of credit for direct enhancement or to assist a bond issuer in obtaining bond insurance. There are advantages for all parties who use letters of credit. Developers decrease their cost of debt and increase the likelihood of a successful bond issuance. The issuing bank earns significant income from fees (either up front, annually, or in some combination) and does not have to carry the letter of credit on its balance sheet. The creditworthiness of the issuing bank determines the value of the letter's credit enhancement, while the bank ultimately looks to the developer and the project for recourse in the event of default.

Alternative Financing Methods

We here summarize the industry's best alternative financing methods, listed as secondary sources of funding in Table 2–3 and categorized as equity or debt vehicles or credit enhancements.

Equity Financing

Developer financing. One of the most common methods of generating equity is developer financing. Developer financing can also be a win-win proposition for both developer and sponsor and often allows the sponsor to maintain control of the project. However, there can be trade-offs. For example, a known developer provides equity capital to the sponsor to improve the chances of obtaining further loans for the majority of the project; in return, the developer provides necessary services, such as construction, management, or marketing and justifies the minimum or extended return on capital through profits to subsidiaries.

Participating mortgage funds. A participating mortgage fund consists of a guaranteed return plus unguaranteed payments, known as contingent interest or equity kickers, that are paid only if earned. The investor is guaranteed a base rate of return and a share of the cash flow above a stipulated base rate. Investors may also realize a share of the appreciated value of the real estate upon the sale or scheduled refinancing of the project.

Participating mortgage funds typically result in a blended rate, where the investor shares in the upside with a fixed downside, earning perhaps less than the rate on a simple nonparticipating investment fund. In addition, credit enhancements can be used to further reduce the cost of the funds.

Venture capital financing. A less common but increasingly popular method of generating equity for long-term care facilities is venture capital financing. The professional venture capital firms require substantial equity positions and may look for annualized returns above 20 percent for their involvement. Personal guarantees are often required, as is possibly majority control over the development and management of the project.

Consider three defining criteria for venture capital involvement: (1) an experienced management team or management contracts in place; (2) strong project cash flows; and (3) equity participation by the developer or sponsor. Venture capital funds are available for projects from $2 million to $100 million.

Syndication. Since the Tax Reform Act of 1986, syndication has been limited primarily to cash returns on small-issue investments. Before this act, most small-issue real estate syndications, which are not required to register with the Securities and Exchange Commission (SEC), allowed for investor tax-loss deductions, thus lowering investor tax liabilities by offsetting ordinary income. The Tax Reform Act has severely limited pass-through losses, making cash-on-cash returns the focus of present syndication investments.

Investor equity. In securing investor equity, securities laws may apply to any attempt to raise equity in which a person invests in an enterprise and relies on the work of another to generate returns. To determine whether the equity financing of a long-term care project meets with securities laws, consider the following:

- Public transactions involve a large number of passive and unsophisticated investors and one or more active participants.
- Negotiated transactions involve relatively few active and sophisticated participants.
- Private transactions involve a limited number of passive, sophisticated participants and one or more active participants.

To conduct a private offering (thus avoiding the expense and complication of a full-scale public offering), the sponsor must follow SEC rules for private placement exemptions. Six rules, Rules 501 through 506 in the SEC Regulation D Offerings, define the terms and conditions necessary for an exemption from SEC registration:

- Rule 501 defines eight categories of accredited investors that are allowed to finance real estate projects.

- Rule 502 defines the financial requirements of the transaction and prohibits the use of general solicitations or advertising.
- Rule 503 provides for the form of sales notices that may be used in connection with the exemption.
- Rule 504 limits the sale to no more than $1 million during a 12-month period.
- Rule 505 limits the sale to an unlimited number of accredited investors and no more than 35 nonaccredited investors; the rule also permits the sale of no more than $5 million during any 12-month period.
- Rule 506 requires that the sponsor determine the knowledge and investing experience of any nonaccredited investors.

Debt Financing

Taxable bonds. State and local governments as well as borrowers (corporations, partnerships, etc.) may issue taxable bonds. Interest on taxable bonds is higher than on tax-exempt securities, but the coupon rate can be lowered by 150 to 200 basis points by purchasing bond insurance or through credit enhancement.

Taxable bonds do not usually require the reserves of tax-exempt bonds and typically have shorter maturities (10 to 20 years). Taxable securities can be exempt from SEC registration if the issuer is regulated by a government body or if the bonds are insured to decrease their risk to the investor. Bonds sold interstate, however, must be registered in a process that can take up to six months and is governed by the Securities Act of 1933.

The advantages of taxable bonds over their exempt counterparts include the investor's ability to earn arbitrage profits, the sponsor's freedom from state volume caps or restrictions on the use of proceeds, and fewer overall reporting requirements. Sponsors who can identify a cost benefit to issuing taxable bonds should strongly consider privately placing these securities, unless credit enhancements can be secured.

Conventional lending. The numerous sources for conventional lending include banks, savings and loans (S&Ls), insurance companies, and pension and trust funds. The specific type and terms of financing will vary by lender, as will the amount of equity participation. Nonetheless, there are generally two parts to any conventional lending: the construction loan and the permanent financing.

The construction loan is an interim loan that funds the construction or rehabilitation as well as the furnishing and equipping of the facility. The terms of the construction loan usually include: (1) evidence of work in progress, (2) a draw schedule, (3) title evidence and title insurance, and (4) bonds posted by the general contractor or a guaranteed maximum price construction contract.

On completion of the project, a takeout loan is used to pay off the construction loan. The commitment for this permanent financing usually is obtained before the construction loan, although the proceeds are not secured until project completion. Contingencies on permanent financing include: (1) title insurance; (2) a new survey indicating all improvements, boundaries, and easements; (3) a certificate of occupancy and applicable permits and licenses; (4) evidence of utility availability; (5) hazard insurance; and (6) proof of compliance with all federal, state, and local environmental laws.

Foreign capital. Experienced developers with a solid record in the long-term care or real estate market can at times attract foreign capital. Loans are provided in dollar or foreign currency, the risk of which is hedged.

Credit Enhancements

FHA insurance. The FHA will provide FHA insurance for bonds sold to both retail and institutional investors with varying fees typically covering the following: bond counsel, issuer, printing, trustee, rating agency, surety, and underwriter's counsel. Fee structures differ for taxable and tax-exempt bonds.

HUD Sections 221(d)(4) and 213. Under HUD Sections 221(d)(4) and 213, our federal government will insure mortgages made by private lending institutions on cooperative housing projects to be owned or operated by nonprofit cooperative housing corporations. The insured amount may be up to 98 percent of the HUD-estimated replacement cost for Section 213 projects and up to 100 percent for

Section 221(d)(4) projects. The HUD mortgage insurance premium is 0.5 percent.

Although nearly dormant, HUD Section 231 provides between 90 and 100 percent insurance on mortgages for the construction or rehabilitation of housing for the elderly or handicapped. Section 231 insurance commitment allocations decreased from $75 million in 1979 to $2 million in 1983, chiefly for two reasons: (1) lenders prefer the cash payments from Section 221(d)(4) insurance in the event of default and (2) Section 202, as an alternative, provides direct loans at below-market rates for qualified borrowers.

Note that all HUD coinsurance programs were terminated in November 1990. This included coinsurance for Section 221(d)(4), 223(f), and 232 programs, under which the HUD-FHA mortgagee assumed 20 percent of the loan loss if the mortgagee defaulted. Those loans completed under HUD coinsurance, however, continue.

Private mortgage insurance. Developers unable or uninterested in qualifying for HUD or FHA insurance can turn to private companies for private mortgage insurance. For the price of an application fee, up-front fees, and monthly premiums, a developer may be able to enhance the strength of its credit application. The degree of enhancement is directly related to the strength of the insurer.

Federal Savings and Loan Insurance Corporation (FSLIC) insured bonds. A qualified entity may issue FSLIC insured bonds and provide an S&L with cash proceeds. The S&L then issues to the bondholders CDs equal to the bonds' principal amount. The proceeds from the CDs are used to pay the mortgage loan, and the project's sponsor issues a note to the S&L. The CD is considered a deposit on the S&L's books and thus an obligation to be repaid. The bondholder is actually in possession of an FSLIC-backed CD, insured for up to $100,000, regardless of whether the S&L receives the developer's payments. The result is a high-credit rating and low-interest coupon rate.

Corporate surety/guarantee. Through a corporate surety/guarantee, some AAA-rated insurance companies will guarantee the payment of healthcare facility bonds that meet their own specific requirements. For a fee, the developer is able to use the full credit rating of the corporation, usually in tandem with coinsurance.

Guaranteed investment contracts (GICs). A GIC is an agreement between the bond issuer and a third party, usually a bank or insurance company. The contract is a noncollateralized note or annuity contract representing a general obligation of the corporation to pay the debt at a given interest rate on all proceeds available for reinvestment during the term of the contract. The GIC's chief advantages are that it fixes the spread between the cost of the bonds and the current rate of reinvestment and insures the developers against market reinvestment risks during the period of the draw schedule.

Put or option bonds. Although typically issued with maturities of 20 to 30 years, put or option bonds can in portion be redeemed at face value after years 1, 5, and 10. A letter of credit is usually required to issue option bonds, because the issue is equivalent to a balloon payment. The issuer may set up an annual sinking fund to ensure payment of the put or may refinance the project should the bondholders exercise their put option.

Summary

The long-term care and senior living field, although still quite a young industry, holds great promise for developers and sponsors interested in capitalizing on an explosive demographic trend. By creatively pairing equity with credit-enhanced debt and enough capital to demonstrate commitment, developers with strong management teams will find lenders willing to support their projects. Naturally, a sound market study and rigorous financial feasibility analysis will bolster any developer's chances of receiving a warm welcome from the investment community.

For Further Reading

Fox, T. C.; A. P. Frum; J. E. Resende; and Carol Colborn. *Real Estate Development: Long-Term Care and Retirement Facilities.* Matthew Bender & Co., 1990.

Gordon, P. A. *Developing Retirement Communities: Business Strategies, Regulations, and Taxation.* 2d ed. New York: John Wiley & Sons, 1992.

Monroe, S. M., editor. *The Healthcare Investor: For the Senior Care Market.* New Canaan, Connecticut: Irving Levin Associated.

Cleverly, W. O. *Handbook of Healthcare Accounting and Finance.* Vol. 2. Rockville, Maryland: Aspen Publishers, 1989.

CHAPTER 3

Legal Perspectives on Long-Term Care Investing

Christopher J. Donovan

The "extended" long-term care industry recently has become the darling of Wall Street. The industry has enjoyed significant infusions of cash, both for start-up companies and for mature, publicly traded companies. Construction and permanent lenders of all types—commercial banks, real estate investment trusts (REITs), government-sponsored programs, conduits, institutional lenders, and others—have recognized the low default rate of most long-term care operators and finally have begun to penetrate a market niche left void by the real estate development boom of the 1980s. The increasing emphasis for hospitals to discharge patients "quicker and sicker," the emergence of well-capitalized health maintenance organizations, and the broader market forces encouraging downsizing, integration, and efficiency all are making long-term care a more popular investment choice.

Today's long-term care investor has inherited a rather hospitable legal climate as well. A variety of legal structures, adoptable for either investor or debt investment in long-term care facilities and operations, are available. This chapter briefly explores the current legal issues associated with debt, equity, and debt-equity hybrid investments in long-term care. The discussion examines the legal aspects of joint ventures, the sometimes hidden price of venture capital, the consequences of choosing for-profit or not-for-profit status, debt financing issues, regulatory requirements, and the legal roots and blossoms of recent developments in the long-term care industry.

Joint Ventures

The trend to provide integrated networks that can service patients in health maintenance organizations (HMOs) through the entire continuum of care has spawned a flurry of joint venture activity among long-term care facilities and other providers in the healthcare system. These joint ventures have taken several forms. True joint ventures are those in which the equity of both the long-term care provider and another healthcare provider, such as a hospital or home care corporation, are at risk. In other forms of joint ownership and management, the parties carry less risk and have varying degrees of benefit and control. The latter category includes management agreements, leasing arrangements, bedhold arrangements, and other development agreements.

The legal issues associated with certain of these joint ventures, particularly those involving not-for-profit corporations, are diverse. For example, a long-term care provider contemplating a joint venture with a not-for-profit hospital to develop a long-term care facility on or close to its campus would have to structure its transaction around the restrictions placed on the nonprofit hospital's ability to deal with a for-profit entity. This principle prohibits the tax benefits of the not-for-profit from inuring to a private individual.

Antikickback legislation also restricts the ability of a not-for-profit entity to provide benefits to a for-profit facility under certain circumstances. Medicare and Medicaid fraud and abuse, tax-exempt bond financing restrictions, and antitrust concerns also need to be analyzed in such joint ventures.

Due to such complexities and the reluctance of many hospitals to commit significant capital to new construction, many long-term care providers are finding strength in numbers. They are forming affiliations, joint programs, and other legal arrangements with community hospitals that offer access to patients and care for each party—but without the hospital's irrevocable equity participation.

Venture Capital

A significant amount of venture capital, both from individual and institutional investors, has found its way into the long-term care industry lately, fueling the start-up of many new companies. In Massachusetts, in fact, the largest health care venture capital financing in 1994 was for a long-term care provider.

But although start-up money for a long-term care development is readily available in so many situations, the cost may be dear. In return for the infusion of certain equity and/or subordinated debt capital, for example, operators may have to surrender a significant equity position in the company, whether through warrants or other stock interest, in addition to investor board of director positions. The eventual goal of the institutional investor in the start-up is to bring the company through an initial public offering of stock and/or sale to a third party. In this way, the venture capital company can realize a return on its seed capital investment by creating a broader, and possibly public, market for the start-up company's previously private holdings.

More traditional forms of equity investments, such as limited partnership syndications and seller financing, also continue to be available. Sources of such investments come from investment bankers who have customers with high net worth and from institutional investors, such as insurance companies or pension funds, seeking rates of return superior to those available on the stock market.

Choice of Entity

A long-term care provider's choice to establish a for-profit or a not-for-profit entity has significant legal consequences. Laws in some states permit a not-for-profit entity to incorporate without becoming a charitable corporation under the International Revenue Code, Section 501(c)(3). However, most of the tax benefits associated with not-for-profit status, including exemption from income tax and deduction of charitable contributions, require the establishment of a 501(c)(3) corporation. The establishment of a not-for-profit/501(c)(3) corporation may be a precondition to accessing certain financing, such as tax-exempt bonds, or may facilitate the structuring of joint ventures with certain other nonprofit companies.

New forms of ownership, particularly the limited liability corporation, have been enacted in most states. The limited liability corporation structure combines the benefits typically associated with a limited partnership (i.e., pass-through of all tax benefits to the limited partners) and S corporation status (limited liability and control by shareholders).

Additionally, new forms of financing, such as conduit debt securitization, have required multifacility operators to create separate "bankruptcy remote entities." Separate bankruptcy remote subsidiary status for each facility is a requirement of many rating agencies. As a precondition to rating the debt placed on certain facilities (and the lower interest rate it carries), these agencies require that financial difficulties at one facility be insulated from soundly performing facilities owned by the same provider. Typically, bankruptcy remote subsidiaries further require the following:

- Articles of organization and corporate bylaws restricting ownership and operation to the immediate facility.
- Independent or partially independent directors.

Clearly, the choice of entity is not one to be taken lightly or without diligent planning and proper counsel.

Debt Financing

Since most lenders continue to view long-term care financing (whether construction, conversion, or permanent) as an operating loan and not a real estate loan, the fundamentals of good lender underwriting and access to capital for providers will continue to revolve around the feasibility evaluation and appraisal analysis regarding the specific facility sites and operators. Expect lenders, who must manage their own business risks, to be ever vigilant of proposed capital investments in the long-term care field. The strength of the borrower's financial information, the location of a specific project, and the standing of sponsoring entities and/or affiliates will affect the amount of collateral (in addition to the project itself) and equity any lender will require and also will affect whether the loan is to be recourse or nonrecourse to individual borrowers.

In demonstrating an operator's creditworthiness and securing adequate debt financing for a long-term care project, the services of attorneys, valuation and appraisal professionals, feasibility consultants, and financial and strategic planners will prove invaluable. In addition, most loan transactions will require significant legal due diligence regarding the following:

- The property's compliance with applicable land use, environmental, and zoning restrictions.
- Evidence of title ownership.
- Property surveys.
- The compliance of the loan documents with the borrower's charter documents.

Developers will require expert legal opinions to tackle these important issues.

Regulatory Issues

Whether you are an equity investor, lender, acquiror, management company, lessee, or other joint venturer in the long-term care business, you will be subject to the highly regulated long-term care environment. Investors in long-term care have no choice but to obtain or file a number of licenses, certificates, and other documents to comply fully with state and local regulations.

Licenses

Most states require an operator's license for each long-term care facility. Such licenses usually are not transferrable, are site-specific, and must be renewed following annual or biannual surveys.

Most often, a suitability review process precedes the issuance of a license, whether on acquisition of an existing facility or on construction of a new facility. The suitability review accomplishes the following:

- Inquires into the criminal records of all major principals and the provider.
- Determines whether any sanctions have been issued against the proposed applicant by any Medicare or Medicaid certifying parties in the immediate state or in other states.
- Examines the financial capability of the provider to build the proposed facility or program and to provide appropriate long-term care.

Depending on the state, a suitability review can take up to 180 days to complete. Careful providers will retain counsel to thoroughly review local suitability requirements well in advance of any anticipated closing. Providers and their counsel work with regulators to facilitate and, if possible, accelerate the suitability review process.

Letters and Reports

For construction loans, licensure is not a pressing issue, since full licensure typically is not applicable until completion of construction. Still, most lenders will want to receive letters from appropriate regulatory agencies indicating that, as of the date of the construction loan, no evidence of nonsuitability appears. Lenders also routinely obtain certifications from architects that the plans developed for the project are in compliance with current regulatory requirements, even though such requirements may change.

For refinancing of existing long-term care facilities, lenders will also want to see Medicare and Medicaid survey reports showing any substandard conditions, corrective plans and actions taken, current certification status by the Joint Commission on Accreditation of Healthcare Organizations and/or the Commission on Accreditation of Retirement Facilities, and provider agreements with Medicare and Medicaid authorities. The lender's goal here is to confirm that the requisite cash flow is available through Medicare and Medicaid reimbursement dollars.

Certificates

Most states also have certificate-of-need (CON) or determination-of-need (DON) requirements that regulate the number of beds in simultaneous service for a given area or population. Through either process, states control the number of beds and thus the amount of reimbursement coming out of their Medicaid systems.

For most lenders, the major concerns regarding CON and DON approval are:

- The need to have the license transferred on any foreclosure sale to the lender itself or to a new purchaser.
- Any uncertainty surrounding such a transfer.
- The potentiality that the lender will be liable for Medicare or Medicaid overreimbursement to a prior owner.

The latter likelihood is particularly important. If a default has occurred under the loan documents, there is a high probability that the facility has been mismanaged. For this reason, lenders typically require detailed financial reporting requirements to ensure an early warning signal should any financial difficulty arise that may force the lender to take back the project. Lenders also are more apt to apply for a receiver than to move directly to take the facility back and incur the associated lender liability for operating the facility. Savvy lenders also require in their loan documents a right to appoint a management agent in lieu of the current management agent for the facility.

Licensures and certifications also play a role in case of foreclosure. Typically it is the long-term care program operator, whether a tenant or manager, that is licensed—not the owner of the real estate. This is advantageous to lenders because it permits them to foreclose on the real estate and potentially leave the management company in place. The core items of any lending documentation, then, are the various reports, such as cost reports, surveys, and any decertification or other license suspensions, which are required of the borrower for Medicare and Medicaid certification.

Long-Term Care Trends and the Law

The growing dominance of rich managed care organizations, the earlier discharge of acutely ill patients, the market shuffling and shakeout of integration—all of the prevalent business trends in healthcare today have deep roots or resonant consequences in the law.

Integration

The significant infusion of debt and equity capital into the long-term healthcare industry has triggered the proliferation of regional operators as well as the consolidation of larger chains of nursing homes and subacute care facilities. With seemingly unstoppable momentum, bigger players are trampling the regulatory playing field. The mom-and-pop skilled nursing facility may well be a thing of the past. Regional chains today compete to acquire smaller facilities and trade on the black market for DONs in states that have declared moratoriums and other restrictions on the issuance of new DONs or CONs. This maturation of the long-term care industry will also feed into the increasing drive for integration by acute care hospitals and other healthcare providers, such as home care corporations.

Most long-term care facilities will need to plug into these integrated networks and become sophisticated at negotiating managed care contracts with large HMOs. This will inevitably require many long-term care providers to fortify their financial and management information systems infrastructures in order to deal with the managed care industry's outcome protocols and other procedural requirements.

Sophisticated long-term care providers will want to seek out alliances both with patient sources (i.e., HMOs) and with other providers in the healthcare system that can act as significant referral sources.

Subacute Care

Although, as of this writing, subacute care has no specific regulatory definition under federal reimbursement and most state statutes, it has become the hot topic in long-term care circles. Driven by the increasing power of HMOs, healthcare reform, and the attempt to reduce the length of hospital stays, nursing homes have scrambled to upgrade their skilled beds to accommodate patients still requiring acute care after discharge from the hospital—in essence converting to subacute care. Also, large chains have developed outcome-oriented facilities for medically complex or rehabilitation cases.

These subacute care facilities must satisfy all of the state licensure, suitability, DON, and other regulatory requirements for reimbursement set forth above. Although typically associated with long-term care facilities, subacute care sites can be located in a variety of settings, including hospital-based units, skilled nursing centers, assisted living facilities, and long-term care or chronic care hospitals.

Real Estate Mortgage Investment Conduits

A major development in the financing of long-term care facilities over the past two years has been the emergence of the securitization market for debt capital. Commercial banks and other traditional lenders, concerned that overleveraging to a single borrower or overemphasis of one region or industry may place the bank's capital at undue risk, are no longer holding in their own portfolios loans made to long-term care borrowers.

Instead, real estate mortgage investment conduits (REMICs) have been established pursuant to applicable tax laws that permit the origination of loans by REMICs and the sale of underlying mortgage-backed securities to investors through investment bankers. The proceeds of the securities are used as the loan proceeds to prospective borrowers. The debt certificates are tranched by the investment banking company and sold as rated commercial paper, carrying varying degrees of interest depending on the rating provided by rating agencies. The various tranches are blended to form the interest payable on the loan.

REMICs can benefit long-term care borrowers and investors alike. The borrower enjoys competitive access to capital. The investor can more easily obtain a critical mass of assets through a pool of diversified loans (diversified by geographic area, borrower, and industry) that are backed by a rating agency's certifications.

Summary

The future looks bright for debt and equity investment in the long-term care industry. Unchallenged demographic trends that will surely demand brisk facility and program development for nonacute care, long-term care, subacute care, and assisted living services should provide incentives for investors and lenders to provide adequate capital for some time.

In obtaining capital and putting it to work, long-term care developers would do well to acquaint themselves with the finer points of the applicable laws and regulations. The points enumerated here are but clippings of the thicket awaiting the long-term care venturer. The astute investor will save a prominent spot at the conference table for an experienced, expert legal advisor.

CHAPTER 4

Technology and the Home Care Revolution

Kathleen R. Crampton and Stephen B. Kaufman

Healthcare is fast becoming home care. More and more patients today are being discharged from hospitals earlier than years ago and, with the aid of emerging high-technology devices, are receiving comprehensive follow-up care in their own homes. A full range of health conditions that just a few years ago would have required long-term institutional care can now be treated with equal or even greater success in the patient's home.

The home care industry has grown exponentially over the past decade—from small mom-and-pop operations to large, publicly traded national chains. As the industry begins to mature, home healthcare increasingly is being viewed as a commodity, fueling price competition and nationwide consolidation. In this competitive environment, home healthcare agencies are being forced to develop strategies to differentiate themselves by expanding services and reducing costs. Advancements in information systems and in device and telecommunications technology will soon enable home healthcare agencies to integrate a full spectrum of sophisticated home care devices, to provide fresh data for patient monitoring and cost control, and to connect with the emerging healthcare data on the information highway.

As home care proceeds to transform our healthcare delivery system, those who can capitalize on the opportunity to deliver high-touch care streamlined by high-tech applications will enjoy a distinct advantage in the marketplace. No future investment in long-term care will be complete without a firmly established, fully streamlined home care component. And no future investment in home care will be sound without a sound investment in supporting technologies.

The Home Care Industry Today

The contemporary home care industry is driven by its three independent sectors: (1) prescribers, (2) providers, and (3) payers. Physicians prescribe or arrange for patient services by rendering an order to a home healthcare agency. Providers, or home care agencies, deliver services as directed by physicians. Payers—including Medicare, Medicaid, and private insurance companies—reimburse providers for most medical care provided in the home.

Home Care Prescribers

More than 530,000 physicians today practice in over 5,700 general hospitals nationwide.[1] In most cases that require home care services, and usually with the support of hospital discharge planners or social workers, a physician sends written orders to a home healthcare agency requesting specific home care services for his or her patient. For home healthcare agencies to be reimbursed, physician orders must indicate that home care services are medically necessary.

The primary reason physicians use home care is to comply with diagnosis-related group (DRG)-based discharge standards regarding inpatient length of stay. Since physicians remain legally responsible for ordering home care services, they welcome assurances of quality in the home setting.

Home Care Providers

The home care provider base grows daily. As Table 4–1 shows, in 1993 the United States had more than 11,400 home care providers, up from 9,980 in 1992. The 30 largest agencies, almost all of which are for-profit, operate more than 2,600 offices nationwide, representing over one-fourth of the home care business.[2] In the past, there has been very little vertical integration in the home care market, with only 1,700 or 15 percent of all home health agencies sponsored by hospitals.[3]

Home care providers can be categorized into five major service sectors: (1) nursing, (2) drug infusion, (3) respiratory therapy, (4) equipment, and (5) rehabilitation. Table 4–2 shows the 1992 and 1993 estimated total revenues and market shares for each of these service sectors.

With such significant growth expected in home care over the next decade, growth is projected for all service sectors as well, with general nursing care expected to grow the fastest. Table 4–3 identifies our nation's major home care agencies by their traditional service strengths; almost all are currently expanding from their major market niches to provide a full range of services.

TABLE 4–1 The Growing Home Care Provider Base: Home Healthcare Offices, 1992–93

	1992		1993	
Provider	No.	%	No.	%
Not-for-profit/VNA	2,074	21%	2,180	19%
Hospital-owned	1,557	16	1,769	15
For-profit	4,935	49	6,157	54
Government	1,198	12	1,125	10
Other	216	2	233	2
Total	9,980	100%	11,464	100%

Source: Marion Merrill Dow, *Managed Care Digest/Long-Term Care Edition,* 1993 and 1994, p. 29.

Table 4–2 U.S. Home Healthcare Revenues and Market Share by Service Sector, 1992–93

	1992		1993	
Service Sector	Revenues ($ billions)	Market Share (%)	Revenues ($ billions)	Market Share (%)
Nursing	$3.80	34.1%	$6.7	40.0%
Infusion	2.90	26.1	3.8	22.4
Respiratory	1.30	11.1	2.2	12.8
Equipment	1.95	17.6	2.5	15.1
Rehabilitation	1.23	11.1	1.6	9.7
Total	$11.18	100.0%	$16.8	100.0%

Source: Bear Stearns, Company Visit Notes: Staff Builders, 27 August 1993, p. 9.

TABLE 4–3 Selected Home Care Companies by Primary Service

Primary Service	Company	Revenues ($ millions)
Nursing	Visiting Nurse Associations (combined agencies)	$1,742
	Olsten Kimberly (home care only)	1,273
	INTERIM	1,060
	Staff Builders	246
Infusion	Caremark	1,783
	Coram (T2/Curaflex/Medisys/HealthInfusion)	475
	Quantum	202
Respiratory	Homedco	470
	Lincare	154
Equipment	Abbey	329
Rehabilitation	Novocare	582

Source: Corporate annual reports.

Home Care Payers

There are five chief sources of payment for home care services: (1) Medicare, (2) Medicaid, (3) private insurance, (4) other payers, and (5) self-paying patients. Table 4–4 shows 1993 home care service expenditures by payer category.

Medicare and Medicaid are far and away the dominant home care payers, covering 55 percent of all expenditures for home care services. Insurance companies, representing 12 percent of the market, include traditional indemnity carriers, health maintenance organizations (HMOs), preferred provider organizations, and self-insured employers. Other payers, representing 12 percent of the market, cover charity, worker's compensation, and donated services. And approximately 21 percent of home care patients are self-payers, covering both copayments and private-duty nursing fees.

Major changes are evolving in home healthcare reimbursement. Medicare intends to replace its current reimbursement system, based on the lower of costs or charges, with per-episode or DRG-like payments. In addition, Medicare continues to encourage Medicare-eligible patients to enroll in HMOs under Medicare risk contracts.

TABLE 4–4 Payer Expenditures on Home Care Services, 1993

Payer	Expenditures ($ billions)	% of Total Healthcare Expenditures
Medicare	$8.1	39%
Medicaid	3.3	16
Insurance	2.5	12
Other	2.5	12
Self-pay	4.3	21
Total	$20.7	100%

Source: Health Care Financing Administration, Office of Actuary, data from the Office of National Health Statistics. See Table 19, *Health Care Financing Review*, 16:1, Fall 1994, p. 291.

Currently, 2.4 million Medicare-eligible individuals, or approximately 7 percent of the Medicare population, are enrolled in HMOs.[4] These managed care developments are forcing providers to maximize productivity rather than volume and to seek the most efficient venue for care. More and more providers can be expected to explore home care as an integral part of their plans to improve the overall reimbursement picture.

The Growing Home Care Market

From 1990 to 1993, annual U.S. spending on home care increased by nearly 87 percent, from $11.1 billion to $20.8 billion, as presented in Table 4–5.

The number of individuals receiving home care has skyrocketed as well. As Table 4–6 shows, in 1992 approximately 3.0 million individuals were discharged by Medicare-certified home care agencies; only one year later that number had climbed to 3.6 million—an increase of 19.8 percent. Significantly, the greatest increases in volume occurred among the non-Medicare-eligible population, those under 65 years of age.

Beyond the very logical proposition that many patients simply have a faster recovery in their own homes, the recent dramatic growth in

TABLE 4-5 U.S. Healthcare Expenditures, 1990-93

	1990	1991	1992	1993
Home healthcare expenditures ($ billions)	$11.1	13.2	16.8	$20.8
% of total healthcare expenditures	1.6%	1.7%	2.0%	2.4%
Average annual increase in home healthcare expenditures	—	19.2%	27.4%	23.8%

Source: Health Care Financing Administration, Office of Actuary, data from the Office of National Health Statistics. See Table 17, *Health Care Financing Review*, 16:1, Fall 1994, p. 281.

TABLE 4-6 Number of Discharges from Medicare—Certified Home Healthcare Agencies

	1992	1993	% Increase
Under 45 years	386,700	530,900	37.3%
45–54 years	143,800	169,900	18.2
55–64 years	237,400	335,300	41.2
65 years and over	2,274,500	2,611,200	14.8
Total	3,042,400	3,647,300	19.9%

Source: National Center for Health Statistics, Advanced Data #235 and #256, Table 5, 1993 and 1994.

home care services can chiefly be attributed to four factors: (1) favorable demographics, (2) aggressive government efforts to control costs, (3) the explosive growth of managed care, and (4) the introduction of high-technology home care equipment. These factors, combined with patient and family preferences for home care, the increasing number of patients with chronic rather than acute conditions, problematic nosocomial infections in the hospital setting, and the expanding and shifting continuum of care, have accounted for the remarkable increase in healthcare services provided in the home.

Favorable Demographics

The nation's burgeoning elderly population provides a rapidly growing market for home healthcare services. The number of persons older than age 65 has increased steadily over the past decade. In 1980 there were 25.5 million people over 65, representing 11.2 percent of the population. By 1990 this age group had increased to 31.2 million, representing 12.5 percent of the population. The over-65 population is expected to grow to 40.1 million by 2010, as a result of the growth in the number of people reaching 65 and increasing life expectancy for those who do.[5]

Persons over age 65 are the largest users of home care services. The elderly have more chronic conditions and are hospitalized far more frequently than the populations in other age cohorts.

Government Cost-Containment Efforts

In 1983 Medicare implemented the prospective payment system (PPS), whereby hospitals are paid a fixed amount for each admission, based on diagnosis, regardless of actual cost. Implementation of this payment system, commonly referred to as the DRG system, has encouraged hospitals to discharge patients as soon as medically possible. Perhaps more than any other single force, Medicare prospective reimbursement and its offspring in the private sector have propelled today's mass migration from traditional inpatient facilities to a variety of outpatient settings, including the patient home. Medicare now pays for a wide range of high-technology services in the home.

Managed Care

Another extremely influential cost-containment phenomenon in recent years has been the proliferation of HMOs,[6] preferred provider organizations (PPOs),[7] and other managed care programs. With over 49.1 million people enrolled in HMOs and 34.1 million in PPOs, approximately 80 million people, or one-third of the total U.S. population, are now enrolled in managed care programs of some sort. These figures show considerable growth from just three years ago, when the penetration rate for HMOs was only 14.7 percent.[8]

Although managed care penetration rates among Medicaid and Medicare programs remain low, this is changing rapidly, particularly for Medicaid programs. A number of states, such as Illinois, have recently passed laws shifting all Medicaid recipients into managed care programs. Greater managed care penetration will foster greater price competition among providers, forcing them to seek the most effective and cost-efficient setting for care. Managed care programs will seek out attractively economic home care alternatives for members who need long-term care.

High Technology in the Home

Until the early 1980s, home healthcare focused on the provision of nursing and custodial services, with procedures such as infusion therapy considered too complicated for the average home care patient. The introduction of smaller, easier-to-use, and computer-aided medical devices greatly expanded the number of procedures that now can be more efficiently and effectively offered in the home. The trend toward the use of increasingly sophisticated medical equipment in the home is expected to continue, thus allowing patients with higher acuity levels to be discharged to home care.

Patient and Family Preference

Given the choice, most patients and their families prefer to have care provided in the home, as long as it can be done safely and efficiently. As the industry redoubles its efforts to adapt the home environment to better serve patients' healthcare needs, more healthcare providers and patients will opt for services in the home care setting.

Chronic Conditions

Thanks to advances in both diagnostic and therapeutic intervention, many conditions once considered to be acute are now considered chronic conditions requiring long-term care. Only a decade ago, patients with certain forms of cancer, for instance, had fairly short life expectancies. With new advances in chemotherapy and other therapies, many patients with cancer now live five to 10 years or more. The corresponding need to treat a growing population of patients with chronic conditions has fueled the demand for services in the home.

Problems of the Hospital Setting

Some hospitalized patients, especially those with particular susceptibilities or compromised immune systems, may succumb to nosocomial infections. Many physicians therefore discharge their patients as soon as possible from a potentially infectious hospital environment into a less populated environment, where the immune system is not as easily threatened.

The Continuum of Care

Payers today are particularly interested in the continuum of care approach, in which a single plan is made to meet patients' healthcare needs before, during, and after a hospital admission. Home care plans today are often developed before admission rather than at the end of a hospital episode. Better care planning has helped increase the demand for home healthcare services.

Impending Changes in the Home Care Industry

Recent, rapid growth is dramatically changing the structure of the home care industry. Pressures are mounting on home care agencies to better do the following:

- Control costs
- Respond to physicians, managed care entities, or case managers.
- Produce reliable patient outcomes data.
- Provide more comprehensive services, while seeking efficiencies through consolidation.
- Develop diagnostic-related programs.

Any home care program's future success will depend on its ability to meet these challenges.

Cost Reduction and Maximization of Resources

Capitation contracting and Medicare's proposed home care PPS system are putting enormous pressures on home care agencies to reduce costs and maximize resources. The successful home healthcare agency will achieve and demonstrate low costs, while maintaining excellent quality and high levels of patient and physician satisfaction.

National Contracting Trends

National employers and insurance companies currently have dozens of home care agency contracts in major geographic markets. To help control administrative costs, employers and insurers are reducing the number of such contracts. National home care chains with broad geographic coverage will enjoy the opportunity to expand their control over significant market segments. Small, local companies not able to bid on national contracts are expected to lose market share to the big chains. To combat this prevailing trend, a number of small, independent agencies are forming alliances for the sole purpose of bidding on national contracts.

Requirements for More and Better Data

Managed care companies demand increasing amounts of data for improved case management and cost control. Home healthcare agencies, which traditionally have been paper-intensive operations, will need to streamline their systems and provide important medical and financial data to managed care entities. Increasingly, agencies are converting to electronic systems to manage intake, scheduling, medical records, and billing information.

Comprehensive Services

Traditionally, home health agencies have focused on relatively straightforward niche markets (nursing, respiratory, infusion, and medical equipment), and so payers could contract with dozens of agencies in one market for several separate services. With the move toward single-provider contracting, agencies must provide "one-stop" services for payers. Many agencies today are expanding to offer payers a full range of nursing, infusion, respiratory, rehabilitation, and durable medical equipment services.

Consolidation

Although the industry remains dominated by a large number of mom-and-pop agencies, consolidation is increasing in response to the growing demand for tighter cost controls. Many new acquisitions and joint ventures have been inspired by the benefits available via economies of scale. In just the past year, Olsten bought Kimberly; Homedco bought Glassrock; and T2, Curaflex, Medisys, and Health Infusion merged to form Coram, a new company. Consolidation is expected to continue thick and fast in the years ahead.

Diagnostic-Related Programs

Home care payers are seeking programs focused on specific types of patients and designed to produce specific outcomes. Home care agencies are developing programs specifically for AIDS, psychiatric, or cardiac patients. Tightly developed medical protocols or care plans are particularly attractive to the case managers who oversee these types of patients. Because they have the ability to bid well-defined aftercare programs with detailed outcomes for specific patients, home care agencies enjoy a distinct advantage in a competitive delivery system.

Opportunities for Improved Care in the Home

As the home care industry has evolved over the past decade, so has our ability to deliver high-quality care in the home. In addition to the introduction of high-tech home care devices, the healthcare services offered in the home have greatly expanded in breadth and scope. With these changes have emerged exciting new opportunities to improve home care.

Improving the Availability of Healthcare Information

However hard nurses and physicians struggle to keep abreast of the health status of their home care patients, new information about each patient is always "a nurse visit away." Changes in medication are too often delayed because vital signs are not monitored concurrently or retrospectively. The lack of such vital data can soon lead to deterioration in a patient's condition and the need for readmission to the hospital, with all the concomitant human and fiscal consequences. Armed with more timely information, healthcare professionals could observe changes in a patient's condition and then intervene appropriately when necessary.

Ensuring Medication Compliance

According to some healthcare providers' estimates, patients take the proper medication only 45 percent of the time. Whether they suffer from despondency, carelessness, or forgetfulness, many noncompliant patients would benefit from a system that provides accurate medication dosages and reminders. Improvements in patient compliance can help lower overall costs to the payer and—better yet—improve clinical outcomes and the patient's quality of life.

Providing Support Services

Weekly nursing visits leave many patients with no system of support for routine medical procedures the other six days of the week. Often, directions left by the nurse on how to change a dressing or clean a catheter are lost or forgotten. Any kind of support services that assist the patient in routine personal and medical care would greatly enhance the effectiveness of home healthcare services.

Maximizing Resources

Our current home care delivery system requires the use of many providers and is expensive. Under a fee-for-service, cost-based structure, home health agencies have had little incentive to improve productivity and have grown dependent on expensive nursing staff. Because home care services must be delivered over a wide geographic area, expensive professionals spend a good portion of their work time sitting in traffic jams or traveling to remote areas.

As per-episode and capitation payments become more common, those home health agencies best able to maximize resources and leverage their nursing resources more effectively will excel. Technological advances such as those described in the next section will be instrumental in home care agencies' ongoing efforts to streamline processes and maximize caregiver resources.

Connecting with the Information Highway

The lack of an automated system linking the home with the home health agency will limit productivity and efficiency of the home care program. Today, basic patient demographic and clinical information commonly is kept both at the agency and in the vehicle of the visiting nurse; rarely is it integrated into physician or hospital charts. Billing information and forms are faxed or mailed to home offices, leaving documents vulnerable to loss or delay. The successful home care agency of the future will need to streamline paperwork and link with emerging community health information networks.

Integrating Independent Devices

A number of high-technology electronic monitoring and therapy products, described in the next section, have become available to the home care patient during the last decade. Many of these products use their own monitoring systems, which then report either independently or through the home healthcare agency to the physician. As stand-alone products, they are not integrated into one system of patient care monitoring. Because of the trends in consolidation and national contracting, systems for integrating this variety of devices will need to be developed if we are to provide care that is at once affordable and effective.

High-Technology Home Care Solutions

The proliferation of high-technology monitoring and therapeutic devices is catapulting us toward new frontiers in home medicine. For every opportunity to improve home care, there is a technology available or in advanced stages of development today. In this section, we list seven prevalent types of high-technology electronic devices currently available to patients in the home.

Personal Emergency Response Systems

Personal emergency response systems (PERS) use a wireless transmitter to activate a receiving base that places a telephone call to a monitoring center. These systems are sold to hospitals and healthcare agencies, which then lease the devices to customers for $30 to $50 per month. The PERS market is currently estimated to be just over $30 million per year and serves an estimated 250,000 patients.[9] Lifeline Corporation, the PERS market leader in the United States, was founded in 1974 and has annual sales of more than $21 million.

Monitoring of Vital Signs

Vital sign monitoring devices continuously or intermittently measure and record patients' blood pressure, pulse, temperature, and heart activity. In the home health arena, a monitor transmits information about vital signs via telephone to a monitoring center provided by the home healthcare agency, physician group, or the company supplying the device. Home health agencies or physician groups buy the monitors and then lease them to patients for $20 to $40 per day. Cardiac Alliance Company markets the Buddy System, a personal health information system for the collection, storage, and transmission of health information gathered at the patient's home; the physician group or home healthcare agency provides the monitoring center.

Single-System Monitoring

Similar to vital sign monitors but focusing on a distinct physical indicator, such as blood glucose or uterine contractions, single-system monitoring devices are designed to assist the patient in collecting data and then transmitting the recorded information by telephone to a central monitoring center, usually sponsored by the device's company. Tokos manufactures a belt device for prenatal fetal monitoring, provided through its own central facility.

Video Monitoring

Telecommunications are moving quickly into the home care market. The technology enabling two-way video conferencing for medical staff has long been available but has not yet been applied in the home care industry. Systems and products currently under development will facilitate video conferencing between home care agency staff and patients in the home. One such system, by American Telecare, provides video cameras and monitors that allow patients to see their caregivers between visits. Video visits enable home care clinicians to detect problems and provide emergency assistance around the clock.

Medication Reminders

As mentioned earlier, noncompliance with medications remains one of the vexing problems among the elderly. Medication reminder systems could go a long way toward alleviating this problem, but Medicare has so far been unwilling to provide reimbursement. In the future, medication reminder and dispensing devices will probably be programmed from a remote healthcare agency, doctor's office, or hospital. CompuMed markets a mechanical device that rolls medications on a rotating belt to dispensing apertures; the device is programmed to follow a patient's prescribed medication schedule.

Infusion Therapy

Over the past 12 years, infusion therapy in the home has mushroomed, becoming the number one business for many home healthcare agencies. The trend in this service segment is toward not only new infusion therapies but also cost containment. For older patients, new infusion therapies include intravenous solutions that can be self-compounded or self-administered by the patient with support from the home care nurse.

Voice-Activated Computer Aids

Providing services such as reading newspapers, dialing and answering the phone, recording appointments, and setting reminders, voice-activated computer aids enable blind and handicapped persons to lead more independent lives. These products can also be used to control appliances in the home with simple voice commands. Since 1989, Transceptor Technologies, Inc., has marketed the Personal Companion, a computer aid that assists patients with appointments and reminders.

Device Networks: Putting It All Together

To date, no serious attempt has been made to consolidate or coordinate the seven types of electronic devices just described. The emerging home care technologies are not single devices but innovative systems that link new and existing devices to better serve their users.

HealthTech Services Corporation has been working to combine state-of-the-art home care technologies on a single chassis and then link entire fleets of integrated devices to a central monitoring station. A home-assisted nursing companion, or HANC, was developed to provide integrated service functionality and networking capabilities. HANC and the HANC Network can be configured to monitor each of the separate electronic systems listed above, with significant cost and efficiency benefits to the user.

Independent home care devices are often monitored by several different home healthcare agencies, introducing conflicting systems to the patient home and bypassing primary care providers and home health agencies in favor of device manufacturers' own monitoring systems. The HANC Network coordinates the full range of home care devices and provides an all-inclusive interface with the primary provider. With all devices linked by the HANC Network, the home care patient sees a single compact unit, and the clinician sees a comprehensive array of monitoring outputs at a glance. The success of the HANC Network and other coordinating mechanisms may well facilitate the next giant leap forward in the development of healthcare technology for tomorrow's home care patient.

Summary

As dramatic growth continues to feed the home healthcare industry, just as surely as it has over the past decade, expect high technology to further streamline home care from a high-touch to an increasingly high-tech enterprise. The challenge for home healthcare providers will be to enhance service delivery by deploying the appropriate new devices, while remaining ever vigilant of patients' need for quality personal care. Intelligently designed and prudently implemented, advancing home care technologies will not supplant but supplement the caring delivery of human medicine.

References

1. *The American Almanac,* Statistical Abstract of the United States, 1993–94 (San Antonio, TX: The Reference Press, 1994) No. 173, p. 119, and No. 179, p. 122.

2. Marion Merrell Dow, *Managed Care Digest, Long-Term Care Edition* (Kansas City, MO: Marion Merrell Dow, 1994) p. 29.
3. Ibid., p. 33.
4. Marion Merrell Dow, *Managed Care Digest, HMO Edition* (Kansas City, MO: Marion Merrell Dow, 1993) p. 31.
5. *The American Almanac,* Statistical Abstract of the United States, 1993–94 (San Antonio, TX: The Reference Press, 1994) No. 47, pp. 17 and 45.
6. Marion Merrell Dow, *Managed Care Digest, Updates Edition* (Kansas City, MO: Marion Merrell Dow, 1994) p. 3.
7. Marion Merrell Dow, *Managed Care Digest, PPO Edition* (Kansas City, MO: Marion Merrell Dow, 1993) p. 11.
8. Marion Merrell Dow, *Managed Care Digest, HMO Edition* (Kansas City, MO: Marion Merrell Dow, 1993) p. 7.

CHAPTER 5

Strategic Facility Planning for Long-Term Care

James C. Morell

Of all the investments a healthcare organization might make in its long-term care future, surely none will prove more vital than the considerable but indispensable investment of time, thought, and resolve. The very scope of such an endeavor, not to mention the costliness of any misstep in today's healthcare marketplace, should give any right-thinking strategist pause.

Still, pursued for good reasons and with adequate preparation, long-term care program development may be just the right strategy for those willing to do their homework first. You may be researching long-term care as a solution to any of the following:

- Place patients in the most appropriate clinical programs within your system to best manage care and cost (a strategy of distribution management).
- Develop a product line that will provide your organization with an additional source of revenue (a strategy of growth management).
- Convert acute care beds to long-term beds (a strategy of capacity management).

Whatever your goal, this chapter is presented with your organization in mind. Aspiring to better manage distribution, growth, or capacity, in fact, represents the most viable strategic rationale for developing a long-term care program.[1] These pages unfold a road map by which to navigate your organization's journey toward a workable long-term care solution. Such a journey typically begins with the strategic commitment to explore the many options for providing long-term care services, and it must continue long after the first patients make a transition into the new program.

Long-Term Care as a Clinical Model

Before taking the single step, it is especially important to appreciate that the commitment to developing a long-term care program is best characterized as a commitment to a clinical model of care rather than to a specific facility design. Although state and federal governments provide various guidelines for long-term care facility development, the required physical/facility adjustments they prescribe are much easier to accomplish than are the organizational transformations required to manage and operate an effective program.

Successful healthcare organizations of the future will view long-term care as an integral part of a clinical model of service delivery and will offer purchasers long-term healthcare as one feature of a managed continuum of services. A clinical model of care provides the strongest foundation for making future operational and facilities decisions; that is, patient care functions best define the operational and facilities requirements for any healthcare organization.

The development of a clinical model of service delivery begins with the aggregation of patients and the analysis of the population to be served, both in whole and in all of its constituent parts. Think of patient aggregation as assembling the "portfolio" of patients whose specific long-term care needs your program will serve. A thoughtfully assembled portfolio of patients enables program planners to tailor the most effective portfolio of services.

Three perspectives of long-term care should be examined within the context of patient aggregation:

- How long-term care services fit on the clinical service continuum.
- The customer's or supplier's perspective.
- The payer's perspective.

For the purposes of our discussion, long-term care customers include patients and families, and suppliers include the individuals directly involved in the care delivery process. Payers are the third parties responsible for financing service delivery through fee-for-service, managed care, or some hybrid payment model.

Long-Term Care's Place on the Continuum

On the clinical services continuum of wellness, assessment, intervention, rehabilitation, and chronic disease management, long-term care patients have traditionally been viewed as individuals requiring extended rehabilitation or ongoing assistance with the management of a chronic disease. Thus, the long-term care patient has long been defined as an individual not sick enough to require hospitalization but too sick to be cared for or treated at home. This approach to patient aggregation provided the market niche for traditional rehabilitation centers and nursing homes.

Today, with advances in technology, increasing financial pressures, and the goal to return individuals to their highest possible level of health status, the greater posthospitalization population is segmented into well-defined cohorts of patients requiring well-defined protocols of care at each step of the care and recovery process. This segmentation has created a system in which patients must make many stops (both figuratively and literally) along the continuum of care. Each stopping point typically requires a rather short stay during which the patient receives a specific treatment or therapy before moving along to the next station on the continuum. Traditionally defined long-term care patients are now being aggregated into groups that might require a series of services at a number of venues, including the delivery of services in the home, over an extended period. Care protocols may call for a specific therapy or treatment to be delivered over a relatively short period—several days to several weeks—before the patient makes the transition to another protocol along a predetermined care pathway. And with each protocol transition, there may be a corresponding change in patient location/environment to better support the care and recovery process.

To identify the operational and facilities requirements for long-term care services, program planners must understand the distinct requirements that must be met at the long-term care patient's every stop along the clinical services continuum.

How Customers and Suppliers View Long-Term Care

Although healthcare customers and suppliers also subscribe to the clinical services continuum of wellness, assessment, intervention, rehabilitation, and chronic disease management, they have a distinctive point-of-view concerning patient aggregation for long-term care. Neither customers (patients and their families) nor suppliers (the caregivers) typically view the rehabilitation segment of the continuum as long-term care. Because each step in the rehabilitation care process brings the patient a step closer to independence and fully functional wellness, customers and suppliers accentuate short-term victories rather than long-term struggles. Neither is the chronic disease segment of the continuum viewed by customers (patients and their family members living with a chronic disease) or suppliers (the team of clinicians who assist in the treatment and management of these individuals and their families) as long-term care. From the perspective of the customer or supplier, chronic disease management is a series of battles won in the war against the disease; long-term care is sought only in the end stages, once the war is all but lost.

To customers and suppliers alike, then, long-term care connotes dependence or even a loss of hope. Rehabilitation care and chronic disease management, on the other hand, constitute therapies and treatments of limited duration and maximum impact that restore or preserve patients' health and independence.

How Payers View Long-Term Care

Health care payers likewise are apt to view long-term care in an unfavorable light—as an outdated model of care and treatment. Ever working to better manage their business risks across the entire continuum of care, payers search for value-added clinical protocols featuring specific goals and predictable outcomes. They restlessly seek out better ways to measure processes and outcomes that will provide the stepping stones to lower costs and reduced risk. And in their desire to ensure patients' speedy return to total wellness and functioning or at least a more stable state, payers worry that the rehabilitation and chronic disease management segments of the clinical services continuum are prone to inefficiencies and waste.

In payers' eyes, then, long-term care often takes too long, and at unacceptably risky terms.

Long-Term Care as a Physical Location

The physical development of long-term care facilities is governed by federal, state, and local regulations designed to ensure access, health and welfare, and fire and safety. The key difference between the physical regulations for long-term care facilities and those for short-stay acute care facilities is the requirement calling for the establishment of residential support (day room, dining, etc.) and therapy (physical, occupational, and recreational) space adjacent to the patient room areas of the unit or center. This section reviews the key issues associated with the development of long-term care capabilities on the main hospital campus within existing facilities, on-campus in new facilities, and off-campus.

On-Campus Development in Existing Facilities

Organizations wishing to develop long-term care capabilities most commonly pursue solutions on the main hospital campus, in existing facilities, especially when beds are already available as a result of low or declining census levels for acute care services. If the organization's strategic direction emphasizes capacity management, conversion of underutilized acute care beds to long-term care service is a viable option. Several space issues should be addressed early in the planning process.

To provide all the residential support and therapy spaces required for essential long-term care services, approximately one-third of existing acute care beds will have to be taken out of service. A 40-bed acute patient care unit potentially becomes a 26-bed long-term care unit, 24 acute care beds convert to 16 long-term care beds, and so on. This space redistribution will likely be necessary whether the existing acute care patient unit is a single-corridor, double-loaded pinwheel, a double-loaded racetrack, or any other design.

Beyond the reduction in bed inventory to accommodate essential service space, the access requirements associated with the Americans with Disabilities Act (ADA) of 1990 and the Rehabilitation Act of 1973, as well as each state's interpretation of these laws, may take additional beds out of service in the conversion to long-term care. This typically occurs when patient rooms designed and sized for two-person occupancy are required to be converted to single-person occupancy to provide adequate space for ADA-rated bathroom facilities. The law requires new or converted facilities to provide handicap-accessible bathrooms, commodes, and shower facilities. A bathroom and commode must be located in each patient room, but showers may be provided in common facilities shared by several rooms.

Another potentially space-consuming factor involves the room size and configuration requirements for conversion to long-term care. Access space around the patient bed and space to accommodate visitors may figure prominently in determining how many existing patient rooms will actually be available for long-term care beds. ADA side-to-side and headwall-to-footwall stipulations for wheelchair access may disqualify some patient rooms altogether, while requiring that some two-person occupancy rooms be converted to single-occupancy rooms for long-term care purposes. Floor space in the patient room for chairs and love seats is at a premium. If adequate space is not available within the existing patient room configurations, then renovations may be required to convert three-patient rooms into two-patient rooms.

The additional requirement that each patient room door must be visible from the nurse station on a long-term care unit (similar to the requirements associated with intensive care units) may take yet more acute care rooms out of the running for conversion. Each unit's floor plan becomes a puzzle that facility planners must solve to satisfy this requirement. Some patient rooms may have to be taken out of service, the existing nurse station expanded or relocated, or a nurse substation developed. If the organization employs a physically decentralized, patient-focused model of care delivery, the administrative centers for each patient care cluster typically occupy rooms throughout the unit in a traditional single-corridor, double-loaded design, offering an opportunity for suitable reconfigurations to meet visibility requirements. Or, if the size and design of the central core in a racetrack floor plan allows, some space may be reallocated to support decentralized administrative centers, and the inventory of patient care rooms may not have to be reduced to satisfy visibility requirements.

In addition to conversion issues that consume space and patient rooms, access to the patient care unit also has to be considered. Access should be reviewed from two perspectives: (1) visitor access and (2) materials support and patient movement. Because long-term care programs are so often residential in character, visitor traffic patterns are different than in traditional hospitals. The long-term care unit may be open 24 hours a day, seven days a week. Family members and close relatives and friends more often are actively involved in the care process, requiring around-the-clock access to the unit. As a result, the unit's main access point should be in close proximity to both parking and an outside entrance to the facility. A second access point to the unit should be dedicated to the movement of patients and materials. Access to the unit for these purposes should be separate from the visitor access point for convenience, safety, and appearance. Separate access, although not required by regulation, further enhances the overall operations and quality of life on the unit.

After satisfactorily addressing each of the issues outlined above, the organization can calculate the resulting size of the proposed unit in terms of the rooms and beds. Based on the patient mix identified to occupy the unit (discussed in the next section of this chapter), planners can rather accurately estimate the number of patient care hours that will be required to serve the future residents of the program. This estimate of patient care hours can then be translated into required staffing mix and hours, which in turn will help determine the number of beds required to make the unit economically viable to operate, assuming there is a market for the program (see the next section of this chapter).

As a rule of thumb, the minimum number of beds required for a long-term care unit is 16, with a maximum of 32. At bed levels above and below this range, unit staffing costs and the overall costs required to support the activities of a long-term care program may become prohibitive. If the range of 16 to 32 beds, with windows, can be accommodated after all of the issues above are addressed, planners can seriously consider converting an acute care unit for long-term care use.

On-Campus Development Requiring New Facilities

The key consideration regarding the development of a long-term care program in a new facility on an existing campus is access. Access considerations for the newly constructed long-term care facility include the following:

- Visitor access should be separate from hospital access. A separate public entry will better guide visitors to the long-term care unit and will further enhance the residential image of the service.

- Access for movement of patients and materials to and from the long-term care unit should also be dedicated and convenient. Most residents of long-term care units require therapies throughout their stays. To receive therapies that cannot be provided on the unit, patients must be transported to and from the therapy area. The routes between the unit and frequently used therapy areas should be direct, with limited traffic volumes. An access point for the movement of food, materials, and reusable products onto and off the unit should also be direct and convenient. The front door should therefore not serve as the entrance for movement of patients and materials.

The facilities-related issues discussed above, regarding floor plan configurations and room and bed requirements for the conversion of existing acute care nursing units, are applicable as well to the development of a new long-term care unit on the hospital campus.

Off-Campus Development

Aside from the cost of the land, the critical step in the development of a freestanding long-term care facility is the selection of an approach to provide effective administrative and support services. Off-campus administrative and support functions can be totally independent of the main hospital campus, totally dependent on it, or a mixture of the two:

- Total independence represents an organization's commitment to provide all long-term care administrative and support services off campus. All office functions, such as patient accounting, business office, medical records, quality assurance, utilization review, human resources, and chaplaincy, are located at the long-term care facility. Likewise, support functions such as dietary, pharmacy, materials management, housekeeping, and maintenance are located within the long-term care facility. With dedicated administration and support capabilities, the freestanding facility operates as an independently owned program rather than as part of a system. As a result, the benefits accrued to

an organization through the facility's system participation would be minimal under this model, regardless of ownership.
- Total dependence represents an organizational commitment to fully use the administrative and support services available on its main campus. Office functions, such as administration, patient care management, and record completion, are located on the long-term care campus to provide direct support to the patient care process. Functions such as patient accounting, business office, and medical records are centralized on the main campus and "purchased" by the freestanding long-term care unit. Thus, space requirements for support services at the long-term care facility are minimal, since main campus services are provided on a just-in-time basis.
- A comfortable midpoint on this continuum might feature the provision of record completion, short-term storage, dietary tray and serving lines, and a pharmacy at the freestanding facility, with food production remaining on the main hospital campus. Other options may be considered based on critical make-or-buy decisions and logistical support systems in place at the time these opportunities are addressed.

The critical question that planners must answer when developing a freestanding facility is how to allocate labor resources. Will hospital therapists be expected to work both campuses, or will the organization be hiring additional therapists for dedicated staffing? The organization must understand the current and future labor pools in the marketplace before making this decision.

Two final notes are in order regarding facilities development for long-term care services:

1. *Consider the future need to expand the long-term program.* Wherever the long-term care program is located—in space converted on an existing acute care unit, on the hospital site in a new space, or on a separate campus—adjacent space and/or land should be identified during the evaluation and planning processes to identify workable solutions for future program development and expansion. Of particular concern is the availability of adequate windowed space for additional

beds and open space for the expansion of therapies and residential functions.
2. *Consider the potential reuse of the facilities being evaluated or constructed for long-term care service.* Given the continuing pressures on both revenues and expenses in the marketplace, flexibility must be built into every capital expenditure. Design solutions for long-term care facilities should provide the organization with a high degree of flexibility for alternative uses of the space at a later date. Flexibility regarding code compliance, space configuration requirements, environmental systems, and accessibility guidelines will be essential. The cost of providing adequate flexibility should be evaluated against the costs required to develop the long-term care program at the present time.

If your organization can at all afford to weave future facility expansion and flexibility requirements into its current facility planning efforts, such investments are most prudently made before the long-term care unit is opened for occupancy.

Determining a Market for Long-Term Care

Just as they regulate the development of long-term care facilities, most states regulate the development of long-term care services. State agencies' market tests typically use franchising methods to ensure the current and future demand for long-term care services. These franchising guidelines—part of the certificate of need, determination of need, or capital expenditure review processes in each state—provide the first threshold that every organization must pass over in the development of a long-term care service. In adhering to state marketplace guidelines, planners must anticipate product line regulatory mandates and must understand the concepts of sheltered services and the integrated service continuum.

Product-Line Regulatory Mandates

The most common guidelines associated with the development of a long-term care service are product line regulatory mandates, which typically segment the marketplace into various levels of care to be provided. By tying market demand to a use rate (patients or patient days per 1,000 population, or beds per 1,000 acute care beds), the state regulates every long-term care product line—from subacute hospital-based care to residential independent living facilities. Every organization is required to use its own state's market demand methods in determining how many beds may be developed for a new long-term care unit. This product line approach to determining service demand represents a capacity management strategy.

Sheltered Services Concept

A variation on these franchising regulations is the concept of a sheltered service or bed. An organization under certain circumstances can develop its own long-term care services or unit to accommodate its own constituents. Specifically, an organization need not demonstrate market demand with use-rate formulas if the new beds are required to serve a specific cohort of patients and will not be competing for patients in the open marketplace. Long-term care beds therefore may be developed to accommodate patients discharged from the organization's hospital or patients affiliated with the hospital's sponsoring organization. Another typical requirement of a sheltered service is that it cannot, when placed in service, take patients away from an existing facility. This stipulation protects the financial viability of existing programs in the marketplace. The sheltered services approach to determining market demand represents either a capacity or distribution management strategy.

Service Continuum through an Integrated Network

Beyond the franchising methods outlined above, which rely on a marketplace driven by a traditional fee-for-service service area, market demand can be assessed from the perspective of an integrated delivery system driven by purchasers seeking high-quality, low-cost solutions for managing patients under a risk contract. This approach takes the concept of sheltered services and beds one step further by assuming the risk for both the patient and the care delivery processes within an integrated network. This concept represents a distribution management strategy.

In examining long-term care as a way to build a full clinical services continuum, an organization should tether its strategic market decisions and supporting modeling techniques to the sheltered services/beds concept or to the greater purposes of an integrated network. Both market models will require the organization to continually rethink how it aggregates patients. As purchasers continue to pressure providers to improve the quality and reduce the cost of care, every long-term care provider will be forced to respond by reaggregating its patients to accommodate a sheltered service concept or an integrated network.

Defining Market Expectations

The marketplace is growing to expect more and more from long-term health care services. Although customers' expectations may not determine whether a program is licensed or launched, they do have a considerable impact on how well an organization can attract and retain patients and maintain the critical support of staff, physicians, employers, and payers. In a highly competitive marketplace, where quality and cost are no longer the key factors in the decision-making process, the overall staying power of a particular long-term care program may be determined chiefly by its ability to meet or exceed the expectations of its various customers: patients and families, caregivers and other providers, and payers and employers. Planners and investors must fully understand the sometimes divergent perspectives of these customer groups.

Patient and Family Expectations

In long-term care facilities, as in most other healthcare settings, patients and their family members expect a comfortable environment, cheerful and clean appearance, pleasant bedside manner, and easy access.

The environment created for the long-term care service or unit is critical to attracting and retaining individual patients as well as managed care contracts. One essential feature is natural lighting and windows that provide pleasing views both in patient rooms and in the residential and therapy areas of the unit. Effective artificial lighting and efficient air handling systems complement the open environment created by natural lighting and views. And finally, the unit's acoustics should suit a residential rather than an acute care service. Keeping the unit quiet is important and helps create the calm, nurturing environment so instrumental to the recovery and rehabilitation process. Within the organization's budget constraints, all of these environmental characteristics should be addressed during the development phase of the unit.

Appearance should be viewed from two perspectives: color palette and housekeeping. The range of colors and textures used for wall coverings, ceilings, window treatments, and floors, as well as pictures, fabrics, and furniture, establish the very character of the unit. Different colors can be especially effective in distinguishing public areas of the unit (the entry, dining areas, recreation and therapy space) from private areas (the patient rooms, staff areas, and support space). The color palette should create a cheerful, open, contemporary image for the unit. From a housekeeping perspective, cleanliness and cleanability are the watchwords. The unit must look clean and be clean at all times. Therefore, material selection and unit design should consider the daily housekeeping requirements of the service.

All individuals who daily come into contact with the patient and family should exhibit a positive, supportive, cheerful, and conversational bedside manner. A bonding between staff members and clients is essential to the quality of long-term caring services. Patients undergoing rehabilitation or recovery from illness and their family members often view themselves as part of the hosting organization's family as well. In cultivating such a high degree of familiarity, patients and families develop a sense of ownership and belonging with the host organization and its caregiving staff. Bedside manner is a key element in the development of this relationship.

Access to the long-term care unit should be viewed from three perspectives: physical access to the unit, patients' and family's access to staff, and timely admission.

Direct physical access through a dedicated entrance is preferable. If this is not possible, the distance and directness of the route from the parking lot to the building entrance of the long-term care unit should be minimal.

Because patient and family participation in care planning and delivery processes greatly affects the overall quality and cost of long-term care services, convenient and frequent access to the unit's caregiving and administrative staffs is essential.

Finally, gaining timely admission to a long-term care unit is a critical issue, especially for the patient's family. Often, families and attending physicians do not enjoy the luxury of time in making arrangements for the appropriate placement of their loved ones and patients. Prompt placement is a highly marketable program feature. Waiting lists, although good for the program, are not at all helpful for patients, families, or attending physicians.

Providers' Expectations

Caregivers and other providers serving the unit share patients' and families' expectations for an environment, appearance, bedside manner, and access conducive to excellent long-term care.

Caregivers and providers expect facilities to reflect their concern for and need to create a residential environment that supports the long-term care program's recovery and rehabilitation focuses. Windows providing natural lighting and pleasing views are important. A feeling of openness is also essential, especially in the residential and therapy areas of the unit. Residential and therapy areas of the unit are best designed as large, open, interconnected spaces. The unit should neither look nor smell institutional. Softer lighting solutions and air handling systems that adequately change the air (at a minimum meeting local codes) will complement the open environment. The acoustical requirements of the unit call for significant sound deadening. As patients work through their individual therapy programs, pushing themselves to the limits of their physical and emotional boundaries, the acoustics of the unit should not broadcast their challenges and successes as if over a public address system.

The caregiver team is conscious of the differences between the traditional acute care setting and the residential environment required for effective long-term care. A homelike unit sets the stage for patient and family participation and ownership in the recovery and rehabilitation processes. Creation of a residential environment that complements the patient care delivery process without compromising the unit's clinical support requirements is a priority for the caregiver team.

The color palette of the unit is also important for the caregivers and the other members of the provider team. Selected colors and textures should differentiate the unit from the more traditional acute care services of the organization. Caregiving providers want the facility to "speak" in warm tones to patients and families, perhaps saying, "Congratulations and welcome! You have arrived successfully at your next stop on the continuum." When finishing the decor, designers can feel free to enhance appearances with attractive materials, since the risks of nosocomial (hospital-based) infections are typically lower on long-term care units. This freedom of choice regarding materials also opens up the color palette, allowing the staff to create a distinctly residential environment that effectively supports good long-term care.

Housekeeping requirements should also be considered from the caregivers' perspective. As the industry moves to more multifunctional staff, the routine maintenance of a clean and organized unit will become the shared responsibility of all staff assigned to the service. Design features that facilitate easy cleanup will be essential. In addition, the selection of materials and colors for high-traffic areas will determine the overall housekeeping requirements of the unit.

What the caregivers really need is a physical and operational setting that frees them to focus their complete attention on patients and families and to perform their jobs efficiently. Adequate staff facilities typically include: (1) charting space, either centralized or in the patient room; (2) conference space for both team conferences and family/team conferences; (3) staff toilet facilities; and (4) storage space for equipment and supplies.

Other highly desirable facility attributes include (1) a close proximity of key areas in order to reduce travel distances; (2) separate traffic patterns between key areas to reduce congestion; (3) therapy and recreation areas separate from the living areas; and (4) dedicated space for families to gather and for the team to meet with families and patients.

Access and traffic control are significant concerns for the caregiver team. Each unit should have a front door especially for families and other visitors. The unit's therapy and recreation areas should be physically clustered near the front door. There should also be direct access from the front door to the living areas and the space for family counseling and teaching. There should be a dedicated entry/exit for patients, staff, and supplies to support a separation of traffic throughout the unit. When access issues are considered, patient safety and security should also be addressed. The staff need to be able to observe their patients at all times, so patients in the therapy and recreation areas need to be in full view of the staff, the doors to the patient rooms must be in full view at all times, and patients and family members moving throughout the unit should be in view of the staff. Many of these access features are regulated or mandated by local or state statutes governing long-term care programs.

Payer and Employer Expectations

Although it may at times seem that payers and employers are interested only in costs and clinical outcomes, certain other market expectations can differentiate one provider from another in a competitive long-term care marketplace. Like patients, families, and caregiver teams, payers and employers do judge long-term care units by their environment, appearance, bedside manner, and access characteristics.

Payers and employers primarily focus on the five senses and on the patient and family candidates' first impressions of the unit. Is the unit appealing to the eye? Does it present well? Are the colors and the lighting bright and cheerful? Do windows offer plentiful light and appealing views? Do the sounds of the unit reflect a quiet, calm environment, nurturing to the healing and recovery processes? Are the sounds of the unit muffled, affording a degree of privacy in a relatively public environment? Does the unit smell clean and fresh without having an aseptic institutional odor? Are all surfaces clean to the touch? Do fabrics complement the color palette, and are they comforting to the touch? Is the food flavorful and tastefully presented? Does the menu offer variety and a balanced diet? (Food service should be appealing to family members also, since family members play an important role in the care delivery process and spend a lot of time on the unit or in the hospital.) Such first impressions determine the marketability of the program and the unit from a purchaser's perspective.

Purchasers' concerns for appearance are closely tied to the environmental characteristics outlined above. The choices of color palette, lighting, and materials used for finishings all play an important role in defining the appearance of the unit. Bright colors, effective lighting, and easy-to-clean materials increase the unit's marketability. Wise selections in these areas make for a solid first impression and help keep operating costs low.

Staff bedside manner is of critical importance to purchasers, as it is their best measure of staff motivation and involvement with the patient and the family. Problems with bedside manner, purchasers know, often reduce the effectiveness of the care delivery process, straining the communication lines between the caregiver team and the patient and family.

Payer and employer concerns regarding access are similar to those presented for the other market segments. Ease of identity and physical access are especially important, as are an efficient layout and design that facilitates circulation and reduces overhead expenditures. Patient, family, and caregiver safety and security is another important consideration for payers and employers. Access is also measured in terms of ease of admission. Like the patient, family, and attending physician, purchasers eschew waiting lists; their long-term care placement needs are immediate.

Identifying Solutions for Long-Term Care

The preceding sections of this chapter have laid a foundation for the development of a long-term care unit. This section examines how health care organizations in the real world have identified solutions for long-term care. Three case studies represent three different perspectives. The first case study describes one organization's development of long-term care as a clinical model, the second reviews the operational development decisions made by a sponsoring institution, and the third focuses on the physical development aspects of the long-term care unit. Although the respective cases address three different key issues, each organization analyzed information and made decisions relative to all three points before implementing its program.

Case Study 1: Clinical Model Development

Having made a solemn commitment to its community to develop a full continuum of services, one institution began developing several off-site subacute care campuses, each dedicated to serving a specific clinical need. One campus serves patients requiring postacute care rehabilitation services as well as patients with chronic diseases requiring various levels of skilled nursing care. For the well patient who may require some degree of assistance, the organization has developed a congregate living center that provides residents with a degree of freedom and independence in a homelike setting. The rehabilitation, skilled nursing, and congregate living programs are supplemented by the organization's home health agency, which provides nursing care and therapies in the home.

Through these investments in long-term care, the organization can now move patients throughout the system to provide care that best matches the clinical status and care requirements of each patient. Also considered is the ability of the family to support the patient in the caregiving process. The programs have succeeded not only in managing the sheltered, existing patients of the system but also in attracting new patients and families to the organization. By providing a postacute care services continuum, the organization has solidified its cost and market position in a highly competitive healthcare community.

Case Study 2: Operational Model Development

In direct response to the problems it faced when discharging acute care patients to local nursing homes, one hospital chose to develop its own long-term subacute care unit. As patient placement became more problematic, so did the organization's financial predicament due to lost reimbursement. Given this recurring but sporadic placement problem, the organization decided to develop a long-term subacute care unit that could operationally and physically accommodate several types of patients at once: (1) patients in need of short-stay, skilled nursing care awaiting transfer to a local nursing home; (2) patients receiving acute rehabilitation services who are not yet ready to return to home or be placed in a long-term rehab facility, and no longer requiring the services of an acute care unit; and (3) acute care patients no longer requiring the intensity of services provided through the traditional acute care unit but not quite prepared to return home to receive home health services.

Operationally, the long-term subacute care unit has become a successful transitional care program that allows the hospital to provide highly effective care for patients who, based on their clinical and home conditions, do not fit into a traditional niche within the provider system. The unit's multiskilled staff provides the service with a high degree of flexibility. The physical aspects of the unit allow it to accommodate a broad range of patients. With such unique operational characteristics, the unit is beginning to attract patients from other organizations, as families and purchasers seek to provide appropriate clinical and operational solutions for their patients.

Case Study 3: Physical Model Development

A third organization already operates an acute care rehabilitation program, skilled nursing facilities, a home health program, and a retirement center. It is planning to convert one of its acute care nursing units into a subacute care unit. The planned subacute care unit will further supplement the current program mix of postacute care services provided by the organization. Presently, the addition of new beds is not an option, so the conversion of existing acute care beds is the only viable physical solution for the organization.

The proposed site of the unit is adjacent to the organization's acute rehabilitation unit, which will allow the two units to share therapy and recreational space through interdependent schedules of activities. The therapy and recreational spaces will be adjacent to each other, allowing for greater staff efficiency, improved supervision of activities, and cross-scheduling of activities. Therapy and recreational functions will be located at the entrance to the unit and in the proximity of the family visitation area and the team/family meeting space, thereby improving family access to these areas.

Furthermore, the unit's location provides direct access to a bank of passenger and service elevators as well as close proximity to the main entrance and lobby of the institution. Access to and from the unit by patients, family members, and the care team is ideal. The unit will have two entrances—one strictly for family and visitors and a second for patients and staff—separating the major traffic patterns as preferred.

Patient rooms will be a mix of single- and double-occupancy rooms with bathrooms. Several of the rooms will be made handicapped accessible, in accordance with ADA requirements for bathroom access and freedom of movement, and will therefore lack private showers. Every patient room and all therapy areas will have views of the rolling hills of the local residential community.

Staff areas will be centralized to provide direct visual observation of patient rooms, therapy areas, recreational space, and the front door.

Implementing the Selected Solution

In each of the three case studies outlined above, the organizations have achieved success by doing their strategic homework before making the commitment to invest in long-term care. This section reviews some of the key organizational decisions that have made these programs success stories in the marketplace today. Specifically, it looks at how the organizations identified criteria to support the decision to develop a long-term care program, how they conducted their make-or-buy analyses, and how long-term care decision processes differed in indemnity and managed markets.

Identifying Decision Criteria

In deciding whether and how to invest in long-term care, all three organizations looked at four key criteria: (1) actual or perceived financial losses associated with acute care patients whose average lengths of stay are too long given their acute care clinical needs; (2) a product-line revenue-generating opportunity in the current marketplace; (3) continuous difficulties in placing acute care patients into other facilities that provide a more appropriate level of clinical services to meet patient needs; and (4) available acute care beds not being used, as a result of a declining demand for acute care beds. Each criterion is discussed below for each case study.

Case study 1. As the major provider of acute care services in its marketplace, this organization has enjoyed being the market share leader. But with such a favored status, in a marketplace then dominated by indemnity model payers, the institution soon found that it was generating a significant number of patients requiring postacute care services. In a cost-plus environment (before the prospective payment system) the organization made money on these postacute patients. However, as the local market moved away from the traditional cost-plus model to the prospective pricing model, extended lengths of stay and additional services became a financial burden. These financial risks were compounded in a local market that failed to generate an adequate infrastructure of postacute care services, making it difficult to place these patients elsewhere. Although postacute patients were receiving appropriate care, the organization was spending more resources in the care process than it was recovering in patient revenues.

An analysis of the situation indicated the organization was generating a large enough population of sheltered patients to make the development of its own long-term care services financially possible. Having acquired a competing hospital just before these cost and placement pressures began to take their toll on financial resources, the organization was forced to close the acquired facility due to declining census levels and continued financial losses from operations. The timing of all these events created the necessary financial incentive, market opportunity, and available physical capacity for this organization to make the investment in long-term care.

Case study 2. The key criterion in this organization's investment decision was the chance to develop a revenue-generating product line to take advantage of a narrow window of opportunity in the marketplace. In addition to this product-line opportunity, the organization saw long-term care services as a way to better manage costs and the movement of its own patients within the healthcare delivery system. As a result, the mix of patients on the unit reflects a blend of outside patients (revenue product line) and sheltered patients (cost management-driven internal referral). To take advantage of these two market opportunities, the organization converted a closed acute care nursing unit into a long-term care unit. As discussed above, unique design features along with multiple staff skills have made this unit an attractive option for many patients in transition between the acute care environment and home care.

Case study 3. Two criteria are driving the planned development of this organization's subacute care unit: (1) the organization is having difficulty in placing patients into long-term care facilities and (2) as a result, is experiencing financial pressures in a marketplace with increasing managed care penetration. The converted subacute care unit would provide an essential link matching clinical requirements (and associated costs) with available revenues (currently discounted and per diem revenues). The organization currently generates a sufficient number of sheltered patients to fill the proposed unit and is capacity bound in its other subacute care facilities. The challenge facing this organization is that there are currently no available beds within the system that could be converted for subacute care. As a result, patients will have to be redistributed throughout the system (a patient aggregation study was conducted) to make room for a nursing unit.

To Make or to Buy?

A rigorous make-or-buy analysis is a key part of every planning process for long-term care services—or for any program or service, for that matter. The question planners must answer is whether the institution should make (convert existing beds or build new beds) its own long-term care capabilities or buy (lease) long-term care capabilities from another provider. Each of the three organizations described in the case studies conducted a thorough make-or-buy analysis. Their assessments and decisions are outlined below.

Case study 1. After assessing the physical characteristics of the recently acquired hospital as well as the physical attributes of the main hospital facility, the organization determined that converting the closed facility to long-term care services was viable. Further, the organization reviewed options for building new space for long-term care services on both campuses as well as on a new, third campus. The marketplace did not offer competing long-term care facilities that could be considered for purchase. On reviewing all of its options, the organization determined that the optimal solution was to make the long-term care unit from existing facilities in combination with new construction. The recently acquired, closed acute care hospital became the likely candidate for conversion. The organization proceeded to demolish sections of the closed hospital, renovate the remaining areas, and add new space to create the necessary environment for long-term care services.

Case study 2. The make-or-buy analysis conducted by this organization entailed an additional complicating factor: mission compatibility. Although bed availability (excess bed capacity) in the marketplace made buying or leasing the service an option, the religious mission of the organization combined with the location and corporate affiliation of the other providers precluded this option in many instances. Where mission compatibility was favorable, bed availability was not. Therefore, this analysis focused on the conversion of existing beds in the system or the development of new beds by the organization. On completing their assessment, planners arrived at an optimal solution that called for the conversion of a closed acute care nursing unit rather than the conversion of existing residential-type beds currently available within the system. In addition, significant weight was placed on the fact that acute care beds were available, and the current acute care census of the organization was going to be the primary patient source for this program. Conversion of other residential beds not located on the acute care campus created other challenges related to patient movement. The construction of new beds was deemed cost-prohibitive given the marketplace, the site, and the availability of existing beds. As a result, an existing acute care unit was converted to a long-term care unit.

Case study 3. Here the make-or-buy analysis started with the consideration of leasing beds from a neighboring hospital. Although the beds were available, the host hospital was not ready to give up its beds. Therefore, the organization began to assess opportunities to make its own long-term care capabilities. Given the state's current position on new beds, planners began to focus on the conversion of existing beds within the system. From this analysis, the need to convert existing acute care beds (with the resulting reaggregation of existing patients due to the closure of an operational acute care bed unit) became the only viable option. Plans were developed to redistribute patients and convert an existing acute care patient unit to long-term care use. Limited consideration was given to converting other existing long-term care beds in the system to this service, as such a conversion would only create additional problems with patient placement and cost management for the organization.

Implementing Long-Term Care for a Managed Care Marketplace

For organizations looking at providing long-term care services as well as for those already experienced in long-term care, the burgeoning managed care market will drive decisions and determine fates. As more local markets shift from the indemnity model of purchaser payment to the managed approach to purchaser payment, so will the associated risks and rewards of providing long-term care shift. Units and programs that once served as product-line revenue generators or cost managers soon become pure cost centers, where the key to continued success is balancing the equation:

$$\text{Quality} + \text{Cost} + \text{Access} = \text{Value}$$

Each of our case studies offers a different perspective of this mounting challenge to provide value in long-term care services.

Case study 1. This organization is well positioned to meet the challenges presented by the managed marketplace. The organization's critical mass of services (reducing operating costs), its overall system capacity to accept and successfully manage large (by current standards) risk contracts, and its geographic location together create an opportunity to make the new long-term care services attractive both to the indemnity market and to the managed care market.

Case study 2. Given the revenue center focus of the organization's investment and the subacute care unit's present configuration, this program may not survive in a managed care environment. Several options will have to be considered as managed care penetrates the market. The unit may have to cater to two distinct purchaser groups—one group representing the covered lives for which the organization is responsible under contract and the other representing patients served in fulfillment of carve-out contracts (lease-type relationships) with other purchasers. In addition, the unit will have to supplement its bed capacity along the clinical continuum to effectively match subacute patient needs with available resources. Additional resources will have to be developed to properly position these supplemental beds and create the critical mass to ensure financial viability at anticipated fixed revenue levels. If these pieces do not come together, leasing the beds or closing the unit will be the only viable options. The organization could ill afford to keep the unit as a loss leader in a managed care environment.

Case study 3. Sponsored by an organization that offers the broadest and most fully integrated service continuum of our three case studies, the proposed subacute care unit could enjoy distinct advantages in the encroaching managed care environment. Still in the planning and design phases, this long-term care unit is being developed with the managed care market firmly in mind, as will more and more long-term care projects in the years ahead. The chief question is whether the organization will be able to balance the value equation in light of pressing capacity management issues.

Monitoring the Solution's Success

Looking toward the future, planners should eye several reliable gauges of the organization's long-term care direction and progress. Whether pursuing a strategy of distribution management or capacity management, your organization should consider certain alternatives and prepare to meet the changing demands of a changing marketplace. This section looks at the key success indicators that organizations should monitor and the contingency plans they should wisely develop.

Configuring Your Decision Dashboard

Key success indicators will vary with the level of managed care penetration in the marketplace. That is, an organization that views its long-term care unit or service as a revenue-generating product line must monitor a different set of gauges than an organization that views its long-term care program as a cost center in a managed environment. For the organization taking both points of view that is concerned about its ability to manage the effective transition from a revenue to a cost perspective, its decision "dashboard" must reflect a blended perspective.

The dashboard for a revenue-focused long-term care unit or service should include the following: (1) occupancy rate; (2) waiting list length; (3) payer mix; (4) net patient revenue; (5) operating expenses; (6) staffing levels and skill mix; (7) admission, discharge, and transfer criteria and protocols; and (8) referral sources.

The dashboard for a cost-focused unit or service should include these indicators: (1) health status of covered lives, with a focus on individuals who may require long-term care; (2) health status and progress of patients on the unit or service; (3) adherence to patient care protocols; (4) individual length of stay on the unit; (5) patient mix; (6) resource consumption per patient day and per covered life; and (7) admission, discharge, and transfer criteria and protocols.

The dashboard for the organization required to manage the long-term care unit or service from both a revenue and a cost perspective should combine all the gauges listed above.

When to Hold and When to Adjust

Organizations that want to develop long-term care as a revenue-generating product line typically fold their hands when operating margins dwindle or when they are no longer willing to maintain a loss leader in the service portfolio. Those wanting to develop long-term care as a cost center typically hold when the resource consumption per patient day or covered life skyrockets and alternative sources for long-term care service are identified.

The real challenges arise in the gray areas created by a shifting marketplace or by a lack of critical mass for the long-term care service. These two interdependent issues are reviewed below.

Shifting marketplace. The shifting marketplace creates the need to (1) revisit the referral sources for the long-term care product line to ensure that they remain accessible and viable; (2) evaluate the potential size of the managed care market available to the organization; and (3) make a commitment to continue accepting open referrals only, shift to a managed patient base only, or maintain a balance of both referred and managed patients. When an organization accepts the challenge of serving both the referral and managed patient bases in the marketplace, it is committing to a cost-center operating concept. This shift in philosophy is essential to the overall success of the long-term care service in a mixed marketplace. Planners should keep their eyes on the dashboard indicators outlined above for the cost center or the combination revenue/cost center model.

Critical mass. If the organization decides to develop the long-term care unit or service as a cost center or a combined revenue/cost center concept of operation, the issue of critical mass comes to the forefront. If the organization is to assume risk for covered lives or accept significant discounts from those purchasers who are assuming that risk, then resource consumption profiles, clinical outcomes, and critical mass questions become the priority issues. Today, 20 to 24 beds is the minimum size for a long-term care unit that strikes the appropriate balance among patient care needs, quality clinical outcomes, resource consumption profiles, and space allocations. Further economies can be achieved with larger decentralized units of 36 to 40 beds. Therapy and residential support components typically occupy 33 percent of a long-term care unit's space, limiting the facility solutions that can be accomplished in available or newly built space. If the resource consumption profile of the unit cannot meet or exceed the expectations of the managed market, then the unit will be competitive only in the open referral market, a segment that is shrinking in most marketplaces.

Organizations should consider adjusting their plans for development of a long-term care unit or service when (1) the unit developed as a revenue-generating product line is not meeting its contribution requirements; (2) current costs associated with the unit, developed as a revenue-generating product line, make it prohibitive to accept managed care patients; or (3) the combined sources of patients—referral and managed—do not generate the critical mass to ensure financial success and high-quality clinical outcomes.

Viable Alternatives

There are several alternatives to consider if at first your long-term care service does not perform well in the marketplace. The first alternative to consider, for a revenue-focused unit, is a change in patient mix. The clinical continuum should be viewed as the foundation for such a decision; the best results typically occur when the unit makes itself available to a contiguous cohort of patients along the continuum. Leaving or creating gaps in the continuum opens the marketplace to competitors, at worst, and generates additional transfer costs for the organization, at best. Building a continuum of care and a niche in the marketplace also enables the organization to increase the size of its patient pool, to diversify its portfolio of purchasers and revenue sources, and to begin altering the cost profile and resource consumption patterns of the unit or service. Niching will allow the organization to sharpen the focus of its marketing efforts. This alternative is designed to expand the revenue base of the unit or service by making the program accessible to a broader base of both referrals and managed care patients.

The second alternative to consider, whether the unit or service is revenue-focused or a cost center, is to rethink the way long-term care is provided and reposition the service in terms of patient base and cost. Such a transformation, typically entailing a change in market focus as well as a change in cost structure, is difficult to manage and secure. However, the potential reward is significant. This alternative is designed to reposition the service and significantly alter its cost structure, thereby enhancing its competitive position.

The third alternative is to divest the unit or service and identify another institution that will provide the service. This alternative relies on a make-or-buy analysis to determine the opportunity costs associated with staying in the business directly or having another organization provide long-term care. The analysis and resulting potential divestiture are especially critical if the service is operated as a cost center in a managed environment. This alternative should also be considered when the host organization's beds are at a premium and there are available beds in the marketplace.

The Future of Long-Term Care

We can learn a lot about the future of long-term care from its present and past roles in our health care system. Historically, long-term care has represented both the best and the worst of what healthcare has to offer. At its best, long-term care has offered a place for patients and family members to seek needed care and support for chronic diseases and extended recoveries that could not be accommodated in acute care settings. At its worst, long-term care has represented the warehousing of patients who didn't fit into the traditional boxes of the delivery system.

These days the long-term care outlook has a decidedly positive spin. Hope for the future is alive and well, thanks to reform efforts under way at the market level as well as the various regulatory reform initiatives at the federal and state levels of government. The next section of this chapter examines the key drivers for long-term care, what the future might hold, and how to build flexibility into a long-term care program.

Key Drivers for Long-Term Care

Every purchaser will always require access to long-term care services as part of a full line of healthcare products. Therefore, every provider must decide how to offer long-term care services that will satisfy its contracting obligations with various purchasers. Specifically, payers will be seeking long-term care services that demonstrate the following:

- The appropriate application of available technologies and clinical protocols that can (1) return the patient to a prior state of health and productivity, thereby reducing the need for additional services and support in the future, or (2) maintain the patient at a stable health status short of full recovery in order to more effectively manage the risks of a reduced health status over an extended period.
- Definable stages of improvement or achieved stability through intense periods of intervention over a relatively short time.
- Appropriate delivery service venues, allowing purchasers to pay for only the resources required to achieve the targeted outcome of the clinical intervention.

In short, purchasers are shaping tomorrow's long-term care marketplace. The better that planners understand purchasers' needs and preferences, the better the outlook for program implementation and success.

Positioning Your Organization for the Future

With purchasers seeking solutions that will reap the best clinical outcomes for their healthcare dollar and thereby maximize the individual and collective health statuses of their covered lives, long-term care seems to be entering a stage of rebirth and redefinition. Strong market players continue to reform our healthcare delivery system. However, the long-term care future will be bright only for those provider organizations demonstrating superior capabilities and value to their purchasers. My advice?

- Identify, recruit, assimilate, and retain the best and most experienced long-term care clinicians and managers. As

ambulatory care has demonstrated to the industry, individuals with impeccable relevant credentials are essential to the success of any long-term care service, regardless of its market niche.
- Offer long-term care directly, as a "made" service, only if your organization can meet or exceed the expectations for value, where value equals quality plus cost and outcomes. If your organization cannot compete by balancing this equation, then identify a partner who can.
- Access to long-term care services will be a highly marketable feature in your organization's future discussions and working relationships with purchasers.
- Remain flexible in your organization's operational and physical configuration of long-term care services. Flexibility will free the organization to develop new clinical protocols and reconfigure its resources as market demands and technology advancements redefine long-term care in the coming years.

This last point warrants further discussion. Organizational flexibility often spells success in the implementation of a long-range strategic facility plan.

How to Build in Flexibility Today

Three critical decisions, addressing facilities, staff, and financial investments, will determine the degree of flexibility that an organization can build into its long-term care service.

From a facilities investment perspective, every effort should be made to develop the space for occupancy by patients with the greatest need for long-term care services. Each regulatory agency that signs off on the development of the facilities and ultimately issues the certificate of occupancy, has developed guidelines for various levels of facilities designed to serve specific classes of patients. These guidelines represent a spectrum that considers the needs of all patients—from those requiring the most assistance to those requiring the least. If the organization plans for and provides facilities that will meet the needs of the patients requiring the greatest level of assistance, the facilities will also be capable of serving patients at the other end of the spectrum. The resulting physical flexibility will allow the organization to change programs in the future without having to make a major investment in facilities.

From a staff development perspective, flexibility may be enhanced through cross-training and multitasking of both clinical and nonclinical staff. A sound investment in staff provides three benefits to the organization. The first benefit is the ability to develop services along the clinical continuum without making significant downstream investments in personnel, especially investments for the release of existing staff and the recruitment of new staff. The second benefit relates to the proven cost savings demonstrated by multiskilled, multitasked staff. The third benefit, stemming from the first two, is the overall stability and continuity of the program's long-term care teams.

From a financial perspective, flexibility is tied to a return-on-investment or return-on-equity period of less than three years, and preferably 18 months. Given the market-driven changes occurring in the industry today, and the demand these changes place on the organization's resources, investments in long-term care, like all investments, should be evaluated as a make-or-buy decision over a relatively short period. The demands of the marketplace, advances in technology, and the sensitivity of the value equation create the need to maintain a short-term perspective on long-term investments. If your organization cannot make the numbers work to meet purchasers' expectations in the short term, then some other provider in your marketplace has probably already met or exceeded their expectations. If this is the case, look at that provider as a strong candidate for partnership.

Closing Observations

As your organization considers its options and opportunities to provide long-term care services, keep three questions at the forefront of the deliberations:

1. Is your organization pursuing a strategy of capacity management or distribution management?
2. Is your organization considering long-term care as a revenue-generating product line or as a cost center in the service continuum required for a managed care market?

3. Is your organization building enough flexibility into its long-term care program to weather the market reform initiatives currently under way in your community?

Your answers to these questions will help you plot the most prudent course for long-term care investment. Organizations that can best clarify their strategic imperatives and anticipate future demands will be in a superior position to offer long-term care services that fully meet or exceed the expectations of the healthcare marketplace today and tomorrow.

Reference

1. Robert, M. *Strategy Pure and Simple: How Winning CEOs Outthink Their Competition.* New York: McGraw-Hill, 1993. (This book illuminates the concepts of distribution, growth, and capacity management.)

CHAPTER 6

The Appraisal of Long-Term Care Facilities

Michael A. Lucas and Janet R. O'Toole

Any investment involving the purchase, sale, development, or financing of property will at some point require an independent appraisal. Simply put, an appraisal is an estimate of value, based on the analyses and opinions of a professional appraiser who has no personal interest in or bias toward the parties involved in the transaction.

The purpose of most healthcare facility appraisals is to estimate the market value of the "going concern" in a proven property operation. Distinct from the value of strictly tangible assets, such as land, buildings, and equipment, the going-concern value of a property includes the incremental value of the business enterprise as well. Going-concern value, then, expresses the total value of a property, including real property, tangible personal property, and property of an intangible nature.

This chapter describes the techniques that professional appraisers employ to determine the going-concern value of existing and proposed long-term care properties. We explore approaches common to all real estate valuations, noting several considerations unique to healthcare facilities and long-term care properties. The methods examined here can be applied to the full spectrum of long-term care facilities and programs.

Establishing the Date of Value

The value of a going concern changes with time. A meaningful appraisal therefore must affix its value estimate to a specific date or to a defined period of operations.

In the appraisal of an existing facility, the value estimate is a snapshot of the property's current or "as-is" market value. Appraisals of proposed properties or additions to existing properties, on the other hand, require estimates of current as well as prospective value, establishing two effective dates of future value: (1) completion of the proposed improvements and (2) stabilization of the business operations. Both current and prospective value conclusions are developed in the context of market conditions as of the date of the appraisal report. Prospective value estimates reflect current expectations and perceptions of market participants along with the factual data available as of the report date.

A long-term care facility is usually planned and staffed to operate at a stabilized occupancy of 85 to 95 percent of available beds. The projected time period from completion of construction to stabilized occupancy is a critical factor in determining both the feasibility and the market value of a proposed project. Called the "absorption period," this projection window is usually based on a monthly growth in a facility's average daily census.

The appraisers must either be provided with a detailed market study prepared by others or conduct a thorough market analysis as part of their own assignment. A market analysis gauges the existing and potential supply-demand relationships that affect the subject and competing properties within a defined submarket. Regardless of who prepares the market study, the appraisers must be able to support the assumptions and conclusions used to project the future income stream of the facility.

Three Valuation Methods

Appraisers estimate the market value of a healthcare facility using three principal valuation approaches: (1) the cost approach, (2) the sales comparison approach, and (3) the income approach. For long-term care facilities, the income approach generally provides the single best indicator of market value, but the purpose and intended use of the appraisal will dictate the possible inclusion of one or more approaches. To best serve their clients, appraisers usually incorporate all three approaches in arriving at their final opinion of market value.

Each approach offers unique perspectives and advantages. An investor embarking on a facility acquisition, for instance, needs to know the value of the individual assets included in the purchase. Only the cost approach separates land, building improvements, site improvements, intangible assets, and personal property—highly useful information that helps planners establish the depreciable basis for federal income tax reporting, provide support for real estate tax protests, and determine the facility's Medicaid and Medicare reimbursement rates.

Cost Approach

The basic premise of the cost approach is that no knowledgeable buyer will pay more for an existing facility than the cost to construct a similar, new facility. Like the other approaches, the cost approach is based on comparison. Here, the cost to develop a healthcare property is weighed against the value of an existing healthcare property. Essentially, the appraiser compares the value of the subject facility to the value of a newly constructed facility with optimal layout, design, and materials. Also, the Uniform Standards of Professional Appraisal Practice state (under Standard Rule 1-2) that the appraiser must "identify and consider the effect on value of any personal property, trade fixtures, or intangible items that are not real property but are included in the appraisal."

In performing an appraisal by the cost approach, the appraiser estimates costs assuming one of two scenarios. The first scenario includes the cost to build a reproduction (an exact replica) of the existing facility as of the date of value. In this cost estimate, it is assumed that the building would be built with the same construction techniques, materials, and design and would contain all the deficiencies of the subject. The alternative scenario, and the one used most often, includes the cost to build a replacement facility (a building of equal utility) employing up-to-date construction techniques, materials, and design.

Whether the reproduction or the replacement scenario is used, the cost to build a new facility is estimated. Some appraisers will rely on national cost figures or seek the services of professional cost estimators. The cost estimate includes all costs to construct the building: hard costs, soft costs, and an incentive to induce the entrepreneur to take the risk of developing the property. Other useful sources of replacement cost data include actual long-term care projects being planned and constructed in the market area. Similar projects can be summarized in project cost comparables, as presented in Table 6–1. Once the replacement cost is established, accrued depreciation is deducted.

TABLE 6-1 Sample Project Cost Comparables

Item	Costs	Cost per sq ft	Cost per bed
Project 1			
Site	$ 843,000	$ 10.04	$ 4,215
Direct construction	6,248,000	74.44	31,240
Indirect construction	1,300,000	15.49	6,500
Furniture, fixtures, and equipment (FF&E)	350,000	4.17	1,750
Total project costs	$8,741,000	$104.14	$43,705

Comments: The facility is a new two-story, 200-bed, masonry and flexicore building. The gross building area is 83,932 square feet (including basement area). This project opened in August and had an ADC of 20 patients within one month. The owner-developer has about six other facilities in the area, all built by a related-party construction company and all very similar. This breakdown does not include any developer's profit. Also note the low cost per bed for equipment and furnishings.

Item	Costs	Cost per sq ft	Cost per bed
Project 2			
Site	$ 1,100,000	$ 12.71	$ 5,500
Direct construction costs	7,513,000	86.78	37,565
Indirect construction costs	1,931,000	22.30	9,656
FF&E	1,288,000	14.88	6,440
Total project costs	$11,832,000	$136.66	$59,160

Comments: This facility is a new three-story, 200-bed, brick and steel frame building. Gross building area is 86,580 square feet (including basement area). This project opened in June. Unique and luxury class, it caters to the Jewish population and has a second, Kosher, kitchen and an area for worship. These costs include the developer's profits (included in soft costs). Also note the high cost per bed for equipment and furnishings.

Accrued depreciation may be due to any of three causes:

1. *Physical deterioration,* which measures the wear and tear on the property over time.
2. *Functional obsolescence,* which is caused by less-than-optimum utility in the layout, materials, etc.
3. *Economic obsolescence,* which is caused by factors outside the property itself, such as a depressed market or neighborhood deterioration. The total accrued depreciation is then subtracted from the replacement cost.

Last, the depreciated cost of the furniture, fixtures, and equipment (FF&E) is estimated; land value and the depreciated cost of site improvements (fences, paving, etc.) are determined; and these three figures are added to the depreciated cost. Thus, the cost approach, sometimes also called the "builder's method," proceeds in seven steps:

1. Estimate the cost to replace the building.
2. Deduct accrued depreciation.
3. Estimate the cost to replace the site improvements.
4. Deduct the depreciation accruing to the site improvements.
5. Add the depreciated value of FF&E.
6. Add the land value as vacant and ready for development, as estimated from comparable sales.
7. Calculate the total value of steps 1 through 6.

Table 6–2 presents a sample cost approach summary highlighting the outputs of these seven steps.

The cost approach is considered most appropriate when valuing new, proposed, or relatively new construction and when little accrued depreciation is present. It is also valuable in estimating the value of proposed additions and renovations. It may be the only method for estimating the value of special-use properties (such as magnetic resonance imaging facilities or surgery centers) that are not frequently

TABLE 6–2 Sample Cost Approach Summary

Replacement cost	$100,000
Less accrued depreciation	
Physical deterioration	− 1,000
Functional obsolescence	− 2,000
Economic obsolescence	− 0
Depreciated cost of the building	97,000
Depreciated cost of site improvements	+ 10,000
Depreciated cost of the building and site improvements	107,000
Depreciated value of FF&E	+ 30,000
Estimated land value	+ 20,000
Depreciated replacement cost of the property	$157,000

bought and sold. One limitation to the cost approach is that it does not contain the going-concern value but only includes the value of the tangible assets: land, buildings, site improvements, and FF&E.

In the appraisal of healthcare properties, however, the cost approach is especially useful when a separate valuation of land, buildings, site improvements, and intangible business value (or going-concern value) is necessary. Well-functioning nursing homes, continuing care retirement communities, hospitals, and retirement living centers contain substantial intangible value, such as goodwill, patient lists, waiting lists, reputation, assembled workforce, accounting systems, management, certificate of need (CON), and licenses.

When the final value (including intangibles) is determined, the difference between the value calculated by the cost approach and the final value equals the worth of the intangible assets. Therefore, the intangible assets are a residual value.

Sales Comparison Approach

The premise of the sales comparison approach is that the market value of a property can in part be determined by comparing the subject property to similar properties that have been sold recently or that are listed for sale. According to this theory, the market value of a property can be determined by adjusting the prices of comparable or similar properties. This approach is most valuable when a number of very similar properties located near the subject have sold recently, and when the property is a fairly simple one, such as a home, an apartment building, or an industrial building. The sales comparison approach usually is not applicable to special-purpose properties, since few comparable sales would be available for analysis.

In the sales comparison approach, the appraiser considers the similarities and differences of the subject property and the comparable sales properties that are most like the subject and most representative of market trends. The appraiser reduces the comparable properties' sales prices to a common denominator that allows like comparison of facilities, one to the other and to the subject. For healthcare properties, the typical units of comparison are price per bed and price per square foot. Table 6–3 displays the data that appraisers typically collect relating to the sale of a long-term care facility.

TABLE 6-3 Comparable Improved Sales Data

General sales data

Facility	
Location	
Granter	
Grantee	
Date of sale	June 1989
Sale price	$2,070,000
Number of beds	69

Property data

Land size	4.09 acres
Building area	26,878 square feet, plus 10,262 square feet of basement area
Building area per bed	390 square feet
Year built	1965

Operating data

Occupancy percentage	96.0%
Private pay percentage	86.8%
Revenues	$1,748,989
Operating expenses	$1,499,886
Operating expense ratio	86%
Net operating income	$249,103

Market value indicators

Price per bed	$30,000
Price per square foot	$77.02
Gross income multiplier	1.18
Overall capitalization rate	12.0%
Comments	This property is located in a fully developed neighborhood of good-quality, single-family residences. This facility was in good overall condition when sold.

Once the common denominator is figured, the appraiser analyzes at least nine elements of comparison or contrast:

1. *Property rights conveyed.* The appraiser must make sure the property rights conveyed in the sales of the properties and those conveyed in the subject facility are the same, or adjustments must be made. For a nursing home, for instance, an appraiser valuing the fee-simple rights (essentially the 100 percent unlimited rights) in the subject property must be

sure that the fee-simple rights—as opposed to the leased-fee value (the value of the property with an existing lease)—were likewise conveyed in the selected comparable sales.
2. *Financing terms.* When buyers obtain below-market financing, they may have paid a higher price for the property to obtain such favorable financing. If the mortgage interest rate is above market, they may have paid a lower sales price. Knowledgeable appraisers adjust the sales price to be "cash-equivalent."
3. *Conditions of sale.* The motivations of the buyer and the seller must be investigated because they often influence the sales price. For example, sellers who need immediate cash may sell a property at a below-market price, or related parties may sell to one another (say to family members or related corporate entities) at nonmarket prices. If the appraisers find that a sale was not an "arm's-length" transaction, they must make appropriate adjustments.
4. *Market conditions.* The most typical adjustments are made to compensate for any changes in market conditions that have occurred since the dates of sale for the comparables and the effective date of value for the subject. Loosely speaking, this is often referred to as a "time" adjustment. Values may have appreciated or depreciated, and the appraiser must make adjustments to compensate for such changes in the marketplace.
5. *Location.* If any comparable sale property has different locational characteristics than the subject, an adjustment, typically expressed as a percentage, will need to be made. A facility on or near an arterial street with excellent access, for example, may have a location superior to one located on a residential secondary street or in a less desirable neighborhood.
6. *Physical characteristics.* Adjustments for physical characteristics, in addition to the market conditions and location adjustments, are the most familiar to the informed layperson. For any physical characteristics, such as building size, that distinguish the subject from any of the comparable sales, adjustments must be made. Examples of such physical

characteristics include the quality of construction and building materials, age, condition, amenities, and functional utility. For a nursing home or retirement center, distinguishing physical characteristics may include the overall square footage of the building. If the square footage of the subject property is 450 square feet per bed, for example, and the comparable facility measures 300 square feet per bed, the price per bed of the comparable sale should be adjusted upward to show that the subject property is more valuable, since it probably has not only larger patient rooms but more and larger common areas and wider hallways.

7. *Economic characteristics.* Factors that affect revenue become most important in the appraisal of an income-producing property. Important points of comparison in healthcare properties can include the operating expense ratio (total expenses divided by the effective gross income), which indicates the efficiency of a property, or the ratio of private-paying patients to the total census.

8. *Property use.* If there is a difference in use or highest and best use of a sale and the subject, the appraiser will probably need to make an adjustment. For example, if a hospital is sold and is intended for continuing use as an acute care hospital, it may not be suitably comparable to an acute care hospital that will be converted into a long-term care or substance abuse facility.

9. *Nonrealty components.* For healthcare properties, relevant nonrealty components usually include FF&E as well as the ongoing business value of the enterprise. If the property being appraised contains a full complement of FF&E, it should not be compared with the sale of a building that did not include FF&E, unless appropriate adjustments can be made. We have encountered a situation in which a nursing facility was sold and converted to assisted living; the operating license was not included in this particular transaction, negatively affecting the value of the property.

Other comparison points, such as access to the property, may supplement these nine essential elements.

Table 6–4 outlines the adjustments made to one comparable nursing home sale to estimate the value of the appraisal subject. In a review of the results of this exercise, it is clear that the subject is more valuable than this comparable sale by $5,986 per bed, or 20 percent.

In healthcare appraisals, the problem with the sales comparison approach is that healthcare properties are very complex. Sometimes it becomes quite difficult to determine all the details of a comparable sale (especially income and expenses, which many organizations consider confidential) and to trust the overall magnitude of the many necessary adjustments.

TABLE 6–4 Sample Sales Comparison

Element of Comparison	Percentage Adjustment	Adjustment to Unit Price of Comparable
Sale Price		$30,000 per bed
Adjustment for property rights appraised	+2%	+ 600
Adjusted price		30,600
Adjustment for financing	0	0
Adjusted price		30,600
Adjustment for conditions of sale	+5	1,530
Adjusted price		32,130
Adjustment for market conditions	0	0
Adjusted price		32,130
Adjustments for		
Location	+5	+ 1,607
Physical characteristics		
Size	+3	+ 964
Quality	–2	– 643
Age	+5	+ 1,607
Condition	–4	– 1,285
Building size per bed	–5	– 1,607
Economic characteristics		0
Operating expense ratio	+7	+ 2,249
Ratio of private-paying patients	+3	+ 964
Property use	0	0
Nonrealty components	0	0
Indicated value for the subject		$35,986 per bed

The value determined by the sales comparison approach does, however, contain the value of the going concern.

Income Approach

Healthcare properties are income producing, and investors therefore purchase them for their earning power. The premise of the income approach is that the higher a property's earnings (typically calculated as net operating income), the higher its value. This approach is particularly beneficial in the appraisal of going-concern value—that is, the value of the real estate plus the business value of the long-term care enterprise.

The income approach demands perhaps the most difficult analysis of the three valuation methods reviewed here and, all too often, is not performed well by appraisers who are unfamiliar with healthcare properties.

In using the income approach to appraise the value of a nursing home, for example, the appraiser first looks at the historic income and expenses for the property itself. These figures, when properly organized, will show whether the property has an increasing or decreasing net income, occupancy rate, payer rates (typically for private, Medicare, Medicaid, and other payers, such as the Veterans Administration), and expenses. Since income and expense categories are optimally reduced to a per patient day basis and a percentage of effective gross income basis, the subject property's income and expenses can also be compared with those of other properties the appraiser has valued to determine whether they are appropriate. The appraiser also compares the subject's daily rates and occupancy levels with those of competing long-term care facilities in the market area. Table 6–5 displays the data typically needed to analyze the subject's position in the market.

With an understanding of historic as well as budgeted or forecasted income and expenses, the appraiser can now begin to prepare a stabilized operating statement. In accordance with such a statement, a facility's occupancy and income are deemed stabilized when abnormalities in supply and demand or any transitory conditions stop and when performance levels can safely be predicted to continue over the life of the property. At the most simple level, the stabilized operating statement consists of the items listed in Table 6–6.

Stabilized income is then capitalized into value with the use of one of two methods: direct capitalization or yield capitalization. Direct

TABLE 6–5 Summary of Comparable Facilities and Daily Rates

Comparable Facility No.	No. of Beds	Occupancy Level (%)	Private Patients (%)	1994 Daily Rates (for Private-Paying Patients)	Distance (miles) from Subject
1	127	71%	67%	$75 Semiprivate room	4
2	213	99	5	$62 Semiprivate room $67 Private room	0.5
3	135	90	5	$75 Semiprivate room	5
4	63	92	9	$75 Semiprivate room $100 Private room	6
5	178	90	50	$90 Semiprivate room $115 Private room	0.25
6	189	90	60	$116 Semiprivate room $425–250 Ventilator $105–115 Alzheimer's unit	1
7	304	96	10	$70 Semiprivate room	6
8	81	97	10	$84 Semiprivate room $91 Private room	0.5
9	96	94	15	$95 Semiprivate room	4.75
10	160	98	20	$85 Semiprivate room $95 Private room	5.25
Subject	168	90	46		—

capitalization equals one year's stabilized net operating income (NOI) divided by an appropriate capitalization rate. Yield capitalization calculations include several NOIs over a holding period, as well as the sale price at the end of the holding period, all discounted to present value. Direct capitalization is applicable to properties with steady occupancies and NOIs, for which one year's stabilized net income is representative of the property's income-producing capabilities. If a property is proposed, is under construction, or has not achieved an acceptable—that is, stabilized—occupancy, yield capitalization is preferable because future income potential is better recognized in the calculation of yield capitalization. Such calculations are typically performed with a computer-assisted discounted cash flow (DCF) analysis.

TABLE 6-6 Stabilized Operating Statement (Income Approach)

	Average Daily Census	Total Patient Days	Percentage Utilization	Rate per Patient Day	Total
Income					
Potential gross income					
Private-paying patients		28,269	46.1%	$103.25	$2,918,725
Medicaid patients		22,566	36.8	63.00	1,421,643
Medicare patients		10,486	17.1	205.70	2,156,913
Other third-party payers		0	0	0	0
Total potential gross income	168	61,321	100.0%	$105.96	$6,497,280
Less provision for vacancy	16	5,840	9.5		618,789
Effective room and board income	152	55,481	90.5%	$105.96	$5,878,492
Ancillary revenue				4.87	269,966
Other income adjustments				0	0
Effective gross income				$110.82	$6,148,457

	Percentage of Effective Gross Income	Rate per Patient	Total
Operating Expenses			
Variable expenses			
Ancillary costs	3.4%	$3.80	$210,824
Administration	15.1	16.70	926,516
Nursing	33.1	36.70	2,036,116
Therapy	9.9	11.00	610,280
Dietary	9.0	9.95	552,026
Facility operations	9.2	10.20	565,896
Bad debt	0.7	0.80	44,384
Management fee @ 5% of effective gross income	5.0	5.54	307,423
Total variable expenses	85.4%	$94.69	$5,253,465
Fixed expenses			
Insurance	0.6%	$0.70	$38,836
Property taxes	1.3	1.40	77,672
Total fixed expenses	1.9%	$2.10	$116,508
Capital reserves for replacement @ $200 per bed per year	0.5%	$0.61	$33,600
Total expenses	87.8%	$97.40	$5,403,573
Net operating income attributable to property	12.1%	13.43	744,884
Capitalization rate			11.5%
Indicated value (rounded)			$6,500,000
Per bed			$38,690

Table 6–7 demonstrates the pattern of new patient absorption, in terms of average daily census, for a proposed 100-bed addition to an existing 168-bed long-term care facility. Note that the final stabilized occupancy rate in January 1997 (the beginning of month 19) is 90 percent (242 patients, 268 available beds). Table 6–8 shows the changes in patient mix over the same 18-month absorption period. In this particular case, the actual mix between the 168 existing beds and the new total 268 beds (including the 100-bed addition) shows a slight change. These census and mix assumptions are then input on an interactive spreadsheet that calculates the monthly patient revenue based on the appraiser's estimate of the daily patient rates and their annual growth factors.

The calculation of capitalization and discount rates is a sophisticated process requiring a close knowledge of the capitalization rates shown by sales in the market. The overall capitalization rate used in direct capitalization equals the comparable property's net operating income for the previous year divided by the sale price. All capitalization rates consider rates of return on alternate investments, since real estate is but one of many vehicles competing for investors' dollars. Although complicated to derive and calculate, capitalization and discount rates ultimately reflect the appraiser's best understanding of the investor's perceptions of the quality, quantity, and durability of the income stream.

The income approach requires considerable time, individual attention to detail, and custom design. Our explanation is therefore highly simplified. Again, a professional appraiser with substantial healthcare experience is best qualified to appraise long-term care properties using the income approach and is best suited to explain the approach and its results to clients.

Assignment and Inspection

The appraisal begins when the owner or manager of a property needs to know its value or the value of some portion of the property to secure a construction loan or long-term mortgage financing; to file a real estate tax protest; to purchase, sell, or lease the property; or to support Medicaid or Medicare cost reporting.

TABLE 6-7 Pattern of New Patient Absorption: Average Daily Census (Income Approach)

Year		Month	No. of Patients for Existing 168 Beds				No. of Patients for Proposed 100 Beds				Facility Total
			Private	Medicaid	Medicare	Total	Private	Medicaid	Medicare	Total	
1995	0	June	70	56	26	152	0	0	0	0	152
1995	1	July	70	56	0	126	1	0	30	31	157
	2	August	68	58	0	126	4	0	33	37	163
	3	September	67	59	0	126	7	0	36	43	169
	4	October	66	60	0	126	10	0	39	49	175
	5	November	65	61	0	126	13	0	42	55	181
	6	December	64	61	0	125	17	0	45	62	187
1996	7	January	63	64	0	127	20	0	45	65	192
	8	February	62	67	0	129	23	0	45	68	197
	9	March	62	69	0	131	26	0	45	71	202
	10	April	62	71	0	133	29	0	45	74	207
	11	May	62	73	0	135	32	0	45	77	212
	12	June	62	76	0	138	34	0	45	79	217
1996	13	July	62	79	0	141	36	0	45	81	222
	14	August	62	82	0	144	38	0	45	83	227
	15	September	62	85	0	147	40	0	45	85	232
	16	October	62	88	0	150	42	0	45	87	237
	17	November	62	89	0	151	44	0	45	89	240
	18	December	62	90	0	152	45	0	45	90	242
1997	19	January	62	90	0	152	45	0	45	90	242

TABLE 6-8   Change in Patient Mix (Income Approach)

Year		Month	Private		Medicaid		Medicare		Total ADC
			ADC	Ratio (%)	ADC	Ratio (%)	ADC	Ratio (%)	
1995	0	June	70	46.1%	56	36.8%	26	17.1%	152
1995	1	July	71	45.2	56	35.7	30	19.1	157
	2	August	72	44.2	58	35.6	33	20.2	163
	3	September	74	43.8	59	34.9	36	21.3	169
	4	October	76	43.4	60	34.3	39	22.3	175
	5	November	78	43.1	61	33.7	42	23.2	181
	6	December	81	43.3	61	32.6	45	24.1	187
1996	7	January	83	43.2	64	33.3	45	23.4	192
	8	February	85	43.1	67	34.0	45	22.8	197
	9	March	88	43.6	69	34.2	45	22.3	202
	10	April	91	44.0	71	34.3	45	21.7	207
	11	May	94	44.3	73	34.4	45	21.2	212
	12	June	96	44.2	76	35.0	45	20.7	217
1996	13	July	98	44.1	79	35.6	45	20.3	222
	14	August	100	44.1	82	36.1	45	19.8	227
	15	September	102	44.0	85	36.6	45	19.4	232
	16	October	104	43.9	88	37.1	45	19.0	237
	17	November	106	44.2	89	37.1	45	18.8	240
	18	December	107	44.2	90	37.2	45	18.6	242
1997	19	January	107	44.2	90	37.2	45	18.6	242

Note—ADC indicates average daily census.

The appraiser begins the assignment with an inspection of the existing facility or, for proposed facilities or additions, with a detailed review of construction documents, including architectural and engineering working drawings and the specification book. The appraiser's initial inspection or review establishes a primary understanding of the design, layout, and construction of the physical plant. A good understanding of mechanical and electrical systems is also important. Newly constructed physical plants will in almost every case be designed and built to meet the most current industry standards. But older buildings may have deficient or obsolete systems and components that often are very costly to correct. If no steps are taken to correct such problems, the operating expenses of the facility might be higher than normal.

Gross building size and numbers of beds are the key physical characteristics that appraisers note in their inspections and reviews. In the case of independent living and continuing care retirement facilities, the number of dwelling units is a common term of comparison. From these characteristics, certain useful comparisons and conclusions can be made.

The square foot area per bed, for example, increases as the number of private rooms and ancillary areas increases. A higher square footage per bed usually signals a more desirable facility and a higher value per bed, but it may also mean a less efficient floor plan. Inefficiencies are often found when an addition is made to an existing facility or when an acute care hospital is reused or converted to a long-term care facility. Thus, additional floor area per bed does not always translate into additional value per bed.

Other physical characteristics that the appraiser considers in the valuation of long-term care properties are the number of stories and the adequacy of vertical circulation. Many nursing home operators, for example, find that a one-story layout is preferable. In two-story layouts, elevator capacity may be taxed at mealtimes and an additional dining room may be necessary on the second floor. Also, the original cost of elevators as well as maintenance and operating costs are key considerations.

A building's shape and height are generally a function of the overall size and shape of its site. For long-term care facilities, one very common layout is the double-loaded corridor, with patient rooms lined up along both sides of a single hallway. The proper length of a double-loaded corridor is limited by the distances from patient rooms to exits and nursing stations. Other important features are corridor and door widths.

Much of the physical plant design and various features are regulated by federal, state, and local building codes and by fire and safety codes. The expense of correcting any noncompliance with such codes figures into the appraised value of an existing facility.

Design Features of Long-Term Care Facilities

The special design features and systems of a long-term care facility depend on the kind of business that will operate in the facility. A nursing facility is designed to deliver various healthcare services for patients of specific types, with specific needs. Levels of patient care are classified by intensity, generally measured in terms of hours of nursing care required per day. The general classifications, in increasing order of intensity, are:

- Independent living.
- Assisted living.
- Intermediate nursing care.
- Skilled nursing care.
- Medical specialty unit.
- Specialty subacute care unit.
- Acute care.
- Intensive care unit.

Level-of-care classifications today tend to overlap, as subacute care encroaches on the territory traditionally claimed by acute care hospitals. Many facilities originally designed for independent living are now classified as assisted living facilities; these facilities offer low-intensity personal and healthcare services as their residents continue to "age in place."

Subacute care can be defined as the level between acute and "chronic" care. Acute implies conditions requiring care of limited duration, and chronic implies the persistent need for care over a longer period. Subacute care can be provided in virtually any healthcare setting, including acute care hospitals at the high end of the acuity scale and nursing facilities at the low end. Subacute patients, most commonly defined as requiring more than three hours of nursing care per day, are generally in stable condition. Most nursing home residents require fewer than two hours of nursing care per day and are thus excluded from the subacute category. Most acute care hospital patients require eight or more hours of nursing care per day. Table 6–9 classifies acute, subacute, and long-term care according to daily revenues and nursing hours as well as length of stay.

In today's health care marketplace, investors are keenly interested in estimating the cost to convert acute and traditional long-term care beds to subacute beds. Expect to pay $40,000 to $80,000 to acquire a private-pay nursing bed and about $20,000 to convert that nursing bed to a subacute bed. A thorough appraisal will pinpoint these costs for a particular subject property based on its previous and intended uses.

TABLE 6-9 Level-of-Care Classifications

Level of Care	Revenues per Day	Nursing Hours per Day	Average Stay (days)
Acute	$800	8	5
Subacute	300	3–8	30–60
Long term	100	1–2	200

The Appraiser's Output

Appraisers prepare narrative reports, form reports, evaluations, and consulting assignments. Narrative reports, prepared to provide maximum communication with the reader, are the type most often developed for healthcare properties. Form reports are typically prepared for small residential properties. Evaluations fall into one of two categories: (1) consulting assignments in which no opinion of value is rendered and (2) an appraisal involved in a transaction that is not classified by federal banking agencies as a federally related transaction. A consulting assignment might involve a feasibility study for a proposed property or an alternate use study for an existing property that is not performing well.

All narrative reports contain certain elements, which follow the order of the appraisal process. Table 6–10 presents a typical outline for a narrative report. The report is divided into four parts plus addenda. Although a narrative report must be adapted for each appraisal assignment, our example basically contains all required report elements.

Appraisal Standards and Types

Until as recently as 1987, there were no official appraisal standards. Most appraisers were guided by the Standards of Professional Practice published by the American Institute of Real Estate Appraisers, now the Appraisal Institute. Partly in response to the problems experienced in the savings and loan industry, the Uniform Standards of Professional Appraisal Practice (USPAP) were published in 1987 and have been republished every year since then. These standards provide specific guidelines on what constitutes acceptable appraisal practice

TABLE 6-10 Typical Narrative Report Outline

Part 1: Introduction
Letter of transmittal
Title page
Table of contents
Certification of value
Summary of important conclusions

Part 2: Premises of the Appraisal
Assumptions and limiting conditions
Purpose and use of the appraisal
Definition of value and date of value estimate
Statement as to whether the value estimate is expressed in terms of cash, terms equivalent to cash, or other precisely defined terms
Property rights conveyed
Scope of the appraisal

Part 3: Presentation of Data
Identification of the property and legal description
Identification of any personal property or other items that are not real property
Area, city, neighborhood, and location data
Site data
Description of improvements
Zoning
Taxes and assessment data
History, including prior sales and current offers or listings
Marketability study, if appropriate

Part 4: Analysis of Data and Conclusions
Highest and best use of the land as though vacant
Highest and best use of the property as improved
Land value
Sales comparison approach
Cost approach, including feasibility study, if appropriate
Income approach
Reconciliation of value indications into a final value estimate
Qualifications of the appraiser

Addenda
Detailed legal description, if not included in the presentation of data
Detailed statistical data
Leases or lease summaries
Other appropriate information

Source: The Appraisal of Real Estate. 10th ed. Chicago: Appraisal Institute, 1992, p. 570.

and ethics. Virtually every competent appraiser follows them, and state-licensed or certified appraisers must abide by them.

Another set of standards for federally insured financial institutions came into being in 1989 with the publication of the federal Financial Institutions Reform, Recovery, and Enforcement Act (FIRREA). The FIRREA guidelines were more extensive than USPAP, and many financial institutions and their customers believed that appraisal costs rose and preparation times lengthened as appraisers struggled to meet the stringent FIRREA requirements for additional data and analysis. Accordingly, the Appraisal Standards Board of the Appraisal Foundation (a group representing most major appraisal organizations) defined six types of appraisals that are acceptable to most clients and financial institutions. Appraisers and their clients must agree on what type of appraisal is appropriate for a given assignment, with the shared knowledge that the more complete the appraisal development and reporting, the more reliable the appraisal.

Appraisal Types

In terms of appraisal development, or the extent of work that the appraiser does in data collection and analysis, there are two broad appraisal types: (1) complete and (2) limited.

According to the 1994 edition of the USPAP, a *complete appraisal* is "the act or process of estimating value or an estimate of value performed without the Departure Provision." When the appraiser develops an appraisal without the departure provision (meaning not departing from all the requirements of the USPAP), he or she will use all applicable approaches to value (cost, sales comparison, and income approaches); the value conclusion, therefore, will be arrived at considering all known information about the property being appraised, all market conditions, and all available data.

A *limited appraisal,* again according to the USPAP, is defined as "the act or process of estimating value or an estimate of value performed under and resulting from invoking the Departure Provision." This means that the appraiser and client agree that the appraiser will not perform all three approaches to value or that the value estimate will not consider all known information about the property being appraised, all market conditions, and all available data.

A common example of a departure to which the appraiser and client may agree is the use of only one or two of the three traditional approaches. Other examples would be cases in which the appraiser did not independently verify comparable data or did not inspect the interior of the property. In these cases, the client accepts the risk of reduced reliability in the limited appraisal.

Reporting Options

Reporting options refer to the types, quantities, and qualities of information found in the written report itself. For both appraisal development types (complete and limited appraisals), there are three ways of reporting the appraisal: in self-contained, summary, and restricted reports.

The *self-contained report* presents all of the information pertaining to how the appraiser arrived at a value conclusion and clearly documents all of the information in the report. Any report following the outline in Table 6–5 may be considered a self-contained report. The self-contained appraisal contains, to the fullest extent possible, explanations of the data and analyses used to develop the value estimate and full descriptions of the property appraised, its locale, its market, and its highest and best use. Most consumers of healthcare appraisals would be familiar with this type of narrative report.

The *summary report* digests the information detailed in self-contained report. The document includes summary discussions of the data and analyses as well as summary descriptions of the subject property, its locale, its market, and its highest and best use. This type of report may be less costly than a self-contained report but not to the extent that appraisal consumers might expect, because complete data and analysis not discussed in the summary report must still be retained in the appraisers' work papers.

The third type of report format for appraisal development is the *restricted report*. It is typically a very short document that contains only statements (shorter than summaries) of the data and analyses used in developing the value estimate. As in the case of the summary report, all of the data, reasoning, and analyses touched on in the restricted report must still be retained in the appraisers' work papers.

Figure 6-1 shows a schematic of the two appraisal development processes and the three report types allowable under each. For each

FIGURE 6-1 Appraisal Report Development Processes and Report Types

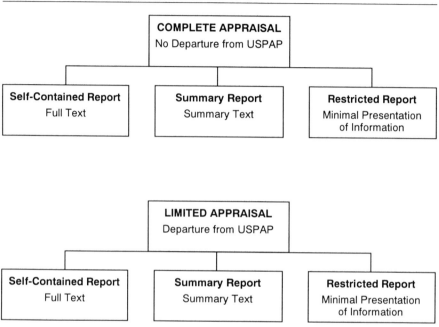

of the reporting types there are 13 reporting requirements that should be included. This ensures that any appraisal will set forth enough information so that it is not deliberately misleading and gives the reader at least a basic understanding of the facts. Clients should be aware that restricted reports, although useful for restricted purposes, include insufficient data for a complete understanding by an intelligent reader and are so short that it is usually impossible to obtain an appraisal review.

Appraiser Qualifications

Until the early 1990s, virtually anyone who wished to do so could call himself or herself a real estate appraiser. As part of the federal government concerns that prompted the USPAP, states were mandated to begin licensing or certifying appraisers.

Licensed appraisers are qualified to appraise single-family homes and small apartment buildings of two to four units. Certified general appraisers are qualified to appraise most other types of large residential, commercial, industrial, and special-purpose properties, depending on their education and experience. An appraisal of healthcare properties requires the services of a certified general appraiser.

The requirements for licensure or certification vary from state to state, but most states require a number of approved appraisal courses; a log of appraisal assignments completed, showing a minimum experience level; and a familiarity with differing property types. Most states also require the submission of actual appraisal reports prepared by the applicant and the passing of a standardized test. Typically there are requirements for continuing education and a requirement that the appraiser conform to USPAP standards, including the Ethics Provisions.

For more complex commercial appraisals, many clients prefer to use a "designated" appraiser. Professional designations are awarded by organizations of real estate appraisers, such as the Appraisal Institute, the Independent Fee Appraisers, and the American Society of Appraisers. The commercial designations of the Appraisal Institute are mainly the MAI, for Member of the Appraisal Institute, and the SRPA, for Senior Real Property Appraiser. The Independent Fee Appraiser's designation is IFA, and the American Society of Appraisers awards an ASA.

All of these designations require more education, experience and experience review, and testing than do the state certifications. Probably the most widely sought designation is the MAI, the most difficult to obtain.

When seeking an appraiser, clients should determine the nature and complexity of their needs, possibly with the assistance of an appraiser. In this way, clients can also decide whether they need a state-certified appraiser or a designated appraiser. Clients should also ascertain the amount of experience the appraiser has had in appraising the specific property type for which the report is required, the appraiser's general experience, and any other qualifications necessary. Knowledgeable clients discuss all aspects of the project with the appraiser and make sure that the product or consulting services proposed are those that most closely fit their needs.

Appraisers' fees vary by project type and complexity, typically ranging from about $4,000 for an existing healthcare facility with an established track record to about $10,000 for a proposed facility with a diverse mix of care levels.

State Regulation and Funding

The value of a long-term care facility is influenced significantly by the regulatory and funding climate of its home state. Obtaining a CON and a license requires careful planning and research and may often reveal information of critical importance to the appraiser's valuation of the property. And since approximately 60 percent of the nation's nursing home beds are utilized by Medicaid recipients, a clear understanding of the state Medicaid program is essential in determining the market value and investment feasibility of any long-term care project.

A state CON program can have a major effect on a facility's existing and future occupancy levels. Each state has developed its own formulas for determining future need for beds in particular regions. Such formulas typically consider existing and estimated regional populations and compare them with existing bed supplies and historical bed utilization.

Certain states have imposed moratoriums on the licensing and construction of new nursing facility beds. In Wisconsin such a moratorium has boosted statewide occupancy levels and created additional demand and higher prices for the purchase of existing licenses. When an existing nursing facility is demolished, downsized, or converted to another use, its licensed beds may be transferred to a new location and included in the expansion or development of a new facility.

In 1987 Wisconsin also established regulations for the state licensing of community-based residential facilities (CBRFs). CBRFs provide assisted living with limited nursing care. The state, primarily looking to building construction, layout, and life safety features, regulates the number of beds, residents' qualifying physical conditions, and staffing at its CBRFs. The state provides no funding for the resident.

The Wisconsin program has been successful in providing certain frail elderly an alternative to the traditional nursing home. Similar residential care or assisted living programs have been established in other states, some of them covered under Medicaid programs. Because residents so often have to "spend down" their assets before being admitted to a nursing facility, the Medicaid ranks are growing by leaps and bounds in many states.

With such regulatory variety among the 50 states, investors would do well to seek out fresh, reliable information from state agencies and associations. Reputable feasibility consultants and appraisers know the lay of the land and expertly guide their clients through what has become a thicket of state requirements for long-term care facilities and programs. The professional appraiser can furthermore express regulatory, licensing, and reimbursement considerations in terms of a property's current or future profitability.

Summary

Professional appraisers today play a leading role in the examination of promising investments in long-term healthcare properties. Experience, knowledge, and independent objectivity together make the appraiser's estimate of value a most reliable piece of the long-term care puzzle. An expertly performed and well-documented appraisal gives all parties to the transaction a solid footing for an informed investment decision.

CHAPTER 7

Information Solutions for the Long-Term Care Provider

Marietta Donne Hance

Long-term care clinicians, administrators, payers, and information systems vendors can expect to spend many years and millions of dollars to reap the benefits of automation. Industry standardization, process redesign, strategic realignment, infrastructure upgrades, employee training, hardware and software development, and testing and implementation of systems all need to occur at individual facilities and at the network level. No small task!

The good news is that many long-term care providers already have started investing. Forty percent of the industry—some 8,000 or 9,000 nursing facilities—will choose software in the next two years to deal with upcoming federal mandates. Another 2,000 to 3,000 facilities will be purchasing software for the second time.[1]

Whether it necessitates the modification of existing systems or the purchase of new equipment and software applications, information solutions must be woven into any successful program design. Considering the prospects of burgeoning automation in long-term healthcare, two chief issues emerge:

1. What are the immediate challenges and benefits of automating in light of current health care dynamics and long-term care operating characteristics?
2. How can a long-term care facility evaluate and implement a feasible information systems solution?

This chapter addresses these questions and provides practical guidelines for long-term care information systems selection and implementation. The first section covers the healthcare industry's impact on information systems, the current state of long-term care and related information systems, and an information systems vision for long-term care. The second section presents specific, practical strategies for selection and implementation of information systems in long-term care.

Ultimately, an organization's return on its investment in computer and information technologies is reaped through productivity improvements and cost reductions at the department, facility, or enterprise level.[2] Simply automating antiquated processes will not yield an acceptable return on investment. In fact, such a misuse of technology can reinforce old ways of thinking and old behavior patterns.[3] Instead, information systems planners should take a hard look at how work actually gets done. Information technology solutions in long-term care must be based on clearly articulated performance improvement objectives and information management needs. Conceived as an integral part of a thorough business process evaluation and redesign effort, new information systems solutions can effectively sustain enhanced business practices.

The Need for New Information Systems Solutions

In formulating appropriate, responsive information technology solutions, long-term care providers must consider their current operations within the context of today's dynamic healthcare industry. Reimbursement and integration are issues that must be considered.

The Reimbursement Landscape

Beginning in 1983, fixed reimbursement based on diagnosis and performed procedures raised the cost-consciousness of all healthcare providers. The emphasis on cost control through improved operational efficiency has accelerated recent trends in hospital downsizing, consolidation, and care delivery in alternative settings. Managed care continues to act as a driving force to redirect care to a limited number of low-cost, high-quality providers. Providers now compete for patients on the basis of cost and quality and are compared with their peers on the basis of cost-effectiveness, clinical outcomes, and satisfaction of consumers (patients, patients' families, and physicians). In response, providers are aggressively redesigning delivery systems to reduce and manage costs and to improve overall performance. There is an ever-sharpening focus on the quantification, management, and allocation of risk.

In this cost- and quality-conscious environment, much attention is focused on improving administrative quality, especially on greater accuracy and speed in billing and collections, the wiser use and coordination of ancillary services, and more effective decision making. Consequently, healthcare organizations are implementing financial performance, operational efficiency, and quality management programs to better evaluate their performance and to learn how to adapt more quickly to meet changing market demands.

Today, a hospital's survival depends on its ability to improve and demonstrate quality, efficiently allocate resources, and maximize productivity. Such profound improvements call for the intelligent collection and dissemination of vital clinical and business data.

Integrated Long-Term Care

Integration of a long-term care facility into a broader healthcare network makes sense. Shorter lengths of hospital stays have led to the establishment of formal referral programs for care after hospitalization. By diversifying into long-term care, many acute care providers hope to offset some of their losses in revenue resulting from decreased lengths of stay.

For the long-term care entity, access to hospital staff expertise, technology, and community image are the chief benefits of affiliation with an acute care provider. Higher-resident acuity and an increasingly complex regulatory environment have created a growing need for medical, allied health (pharmacists, nutritionists, pathologists), and administrative staff in the long-term care setting.[4] Reimbursement requirements, electronic billing standards, and telecommunications advances have further motivated long-term care facilities to seek out hospitals' sophisticated technologies. From a therapeutic perspective, many of the newest diagnostic advances and treatment protocols available in the acute care setting are especially important in dealing with the diseases of the aging.[5] Finally, gaining credibility by association with a well-established hospital or network helps ensure viability for the long-term care partner.

From the acute care provider's perspective, linking to long-term care can solidify the referral pipeline and provide a broader continuum of care, facilitating better control over length of stay and discharge planning. Growing resident populations coupled with limited financial resources in long-term care have created a need for food and nutritional services, clinical pharmacy services, housekeeping, laboratory testing, and laundry services, thus providing an outlet for the acute care facility's excess supply of such services. The potential also exists to establish geriatrics rotations for medical students, to expand geriatric research in medical practice, and to enhance the hospital's community image.

Driving Forces in Long-Term Care

Current long-term care trends, operational characteristics, and regulations underscore the complexity of formulating effective information solutions. An analysis of trends traces important long-range patterns, a review of operational characteristics reveals functional strengths and weaknesses, and a quick lesson in regulatory requirements highlights government's influential role in long-term care.

Long-term care trends. Industry growth, shift in ownership, and providers' response to market changes all play key roles in information systems planning and implementation.

Modest growth. Despite the growing number of elderly persons, growth in the total number of nursing facilities has been conservative because of increased alternative care mechanisms.[6] New opportunities to provide "subacute" care in nursing facilities will depend on future reimbursement dollars available to pay for such services. Lower costs for home care and advances in medical technology that allow care to be provided in the home also explain the modest growth trend in new long-term care facilities. The chief implication for information systems development is that technologies will be implemented largely in existing long-term care structures.

Consolidation. Industry consolidation has snowballed. Multifacility chains that acquire small, independent nursing homes today own one-third of our nation's nursing facilities.[7] From an information systems perspective, this trend presents an opportunity to consolidate and standardize systems within multifacility chains. Standard general ledger and accounts payable functions, for example, could easily serve an entire network.

Shifting ownership. There has been a shift from government-sponsored (public) to proprietary (for-profit) facility ownership. From 1950 through 1991, the government's share in long-term care facility ownership dwindled from 40 percent to 2 percent. During the same period, proprietary ownership more than doubled, from 37 percent to 77 percent.[8] Much of the growth in the proprietary sector is attributable to the 1965 passage of Title XVIII of the Social Security Act, which created the Medicare Program. From an information systems perspective, the trend toward proprietary ownership brings more stringent reimbursement regulations, leading to increased demand for efficient documentation mechanisms.

Changing market. The scope of long-term care is changing. Medical advances that prolong life have increased the number of elderly residents who require more extensive long-term care. Conversely, advances in neonatal medicine have saved the lives of children who previously might have died, creating a younger long-term care resident segment. This trend in the long-term care market supports the development of flexible information systems solutions.

Managed long-term care. There is an increasing managed care presence in long-term care, and strong incentives are at work in the marketplace. Some states have proposed or tested capitated Medicaid payment for comprehensive long-term care services. This trend signals a need for greater cost management in long-term care and further justifies the use of automated information systems.

Networking. Acute care providers are incorporating long-term care services into vertically integrated healthcare networks. There is growing interest among hospitals to establish contractual arrangements with nursing facilities to ensure available beds for Medicare patient discharges. Networks will seek new information systems solutions to help them adapt and extend their service offerings to include long-term care components.

Operational characteristics. The scope of long-term care, admission processes, staffing structures, and financial management can all be improved through better automated information management. But carving new information systems solutions out of an industry beleaguered by government bureaucracy, minimal funding, increasing acuity, and high staff turnover will not be easy. Developing an understanding of the operational characteristics of long-term care will generate ideas for improving efficiency through automation, while defining important practical limitations.

Scope of care. Long-term care can be described as care provided over an extended period to people, regardless of age, with chronic disabilities. Long-term care services are associated with physical or mental disabilities that impair functioning in activities necessary for daily living, such as eating and bathing. Specific care needs are determined by a comprehensive resident assessment. Shorter length of stays in acute care facilities and expanding support for home care suggests that soon only those needing skilled nursing care will reside in nursing facilities, leading to the development of specialty care units for ventilator care, wound care, and intravenous therapy.[9] These changes will alter the case mix and thus the operational approach of the average nursing facility.

Admission process. The long-term care admission process focuses on the prospective resident's needs and the facility's ability to meet those needs. The facility also evaluates the payer status of the candidate. Self-paying residents usually are required to pay in advance. Medicaid recipients must submit proof of eligibility before admission. Medicare eligibility is also considered on admission.

The facility checks its bed availability; if a bed is available, the resident can be admitted. If there is no bed available, the candidate is placed on a waiting list. On admission, an interdisciplinary assessment is conducted by the care team and a care plan is developed. Quarterly assessments are made to evaluate residents' functional status and to determine the appropriate blend of services to provide. The assessments and care plan are updated as necessary and provide the basis for discharge planning, if appropriate.

Staffing. Long-term care staffs typically are divided into three categories: (1) direct care providers, (2) support staff, and (3) administrative staff. Finance systems traditionally have been implemented first, so it is not unusual to find the controller in charge of information systems.[10] Many long-term care facilities use contracted services for rehabilitation (physical, occupational, and speech therapy), pharmacy, lab, and other ancillary services and use volunteers wherever possible. Long-term care facilities have fewer and less experienced business office and management staff than do most acute care facilities. Employee training requirements have increased under state and federal regulations. Resources to upgrade positions are scarce, and it is difficult to attract and retain entry-level employees. Consequently, there is high staff turnover in long-term care.[11]

Financial management. Although the exact figures vary by state, Medicaid and Medicare generally provide more than half of all funding to nursing facilities. Most states limit inpatient stays, but nursing facility stays may be of unlimited duration, which has increased investor interest and government scrutiny in long-term care. The influence of Medicaid is broad. Approximately 42 percent of all nursing facility revenues are paid directly by Medicaid, whereas 65 percent of all nursing facility residents are on Medicaid.[12] Medicaid is administered by the state and financed jointly by the state and federal governments. Overall spending, allowable costs, rate-setting procedures, and covered services vary widely among the states. Nursing facility spending is the largest component of most state Medicaid budgets. Growing concern about annual budget increases at the state and federal levels will lead to increased scrutiny of revenues and operating performance as well as revised eligibility and payment systems to allocate scarce financial resources.

Medicare certification is often a condition for participation in Medicaid. The Medicare program, intended to provide health care services of a more acute, rehabilitative nature, has a limited impact on long-term care. The Medicare program accounts for approximately 5 percent of total revenues for the average nursing facility.[13] Medicare does, however, provide limited long-term care benefits for acute care episodes in the continuum of care. Obtaining Medicare benefits requires a three-day prior hospitalization, a demonstrated need for nursing care and rehabilitation, and facility certification. Even when these requirements are met, Medicare pays for only the first 20 days, and the daily copayment must be made by the beneficiary for the remaining days (80 days for each illness episode).

Prospective case mix payment systems link nursing facility expenses to resident acuity. Accuracy of the case mix data has a direct effect on Medicaid revenues and cash flows and appropriate nursing staffing levels. To receive maximum Medicaid reimbursement and to avoid cash flow delays, long-term care facilities must collect, analyze, and report case mix and financial information in an accurate and timely manner. Although the Health Care Financing Administration is studying prospective payment methods for future implementation, Medicare reimbursement for nursing facilities is still determined on a retrospective basis.[14] To obtain maximum Medicare reimbursement, long-term care facilities must maintain records and collect financial and statistical information in a way that identifies routine and ancillary costs. In the case of a request for exception or reclassification, the burden for proof is on the long-term care provider to demonstrate why costs exceeded the limits.

Regulations. Extensive regulation at all government levels is used to control costs and quality as well as to ensure the fair treatment of all beneficiaries. Resident care revenue streams are derived from the government, third-party payers, private payers (recipients of care or their families), private insurers, and managed care organizations. Long-term care is also subject to significant cost-containment efforts through control of new bed construction, known as the certificate of need process.

Regardless of sponsorship, the long-term care facility must meet certification requirements to receive Medicare and/or Medicaid payments.[15] Compliance with state, federal, and local standards is evaluated by state agencies at least once every year. Nursing facilities must demonstrate their ability to meet requirements and therefore continue to receive Medicaid and/or Medicare reimbursements. Quality of care, quality of life, residents' rights, facility practices, administration, and residents' safety are factored into the evaluation. If the facility is found to be noncompliant, it is required to develop and implement a written plan to correct problems within a designated period.

Unlimited duration of payment for services has stimulated growth in the long-term care sector and increased government oversight.[16] The Omnibus Reconciliation Act of 1987 (OBRA '87) mandated the use of a standardized resident assessment instrument (RAI). The RAI, a uniform system to assess residents' abilities and significant impairments in performing daily functions, includes the following:

- Minimum data sets (MDSs).
- Resident assessment protocols (RAPs)—a structured framework for organizing the assessment to identify problems and plan care.
- Utilization guidelines—state directions about how to use MDSs and RAPs.[17]

Minimum data sets are structured to collect resident information about cognitive functions, communication, hearing and vision patterns, physical functions and structural problems, continence, psychosocial well-being, mood and behavior patterns, pursuit of activities, disease diagnoses and health conditions, nutritional status, oral-dental status, skin condition, medication use, and special treatments and procedures. Individualized care plans are then developed to meet each resident's needs and to ensure that facilities help residents maintain or attain their highest practical level of physical, mental, and psychosocial well-being.[18] Care plans must be updated quarterly.

State and federal licensure and certification hinges on the use of this type of clinical data collection and maintenance system. Mandated on October 1, 1990, the regulations set forth in OBRA '87 were designed to give long-term care facilities the tools to meet regulatory requirements, maximize third-party reimbursement and revenue, minimize claims in suspense and ensure quality and continuity of care.[19] The regulations also were intended to give surveyors a set of standard protocols with which to measure compliance. At first, this type of standardization seemed ideal. But the government also allowed states to enhance the existing MDS, leading to the creation of MDS+. As a result, the MDS/MDS+, related forms, interpretation of compliance, and electronic submission specifications are all state-specific. Each state can submit its own version of MDS/MDS+ to meet licensing requirements imposed by the state. To make matters worse, these requirements do not always match requirements for federal certification for Medicare and Medicaid.[20] Hence, long-term care data collection, coordination, and reporting has become an overwhelmingly complicated endeavor.

Information Investments for the Future

Current long-term care information systems are fragmented and meet only limited needs. Although some healthcare and long-term care providers have made strides in integration, most remain geographically remote, independent, highly specialized healthcare entities. Be it a physician's office, a community hospital, an ambulatory care clinic, or a nursing facility, each entity incorporates its own dialect, protocols, standards, and forms to create a piece of the medical record. A vast array of customized information systems solutions have evolved to meet specific departmental and institutional needs. Proprietary information systems impede information flow across a continuum of care and increase the likelihood of duplication and excessive paperwork. Measurement and interpretation of population outcomes, risk, and customer satisfaction are also difficult.

Barriers against progress. It is not uncommon to find minimal investment in information systems at long-term care facilities.[21] Package systems are less sophisticated than those found in the acute care setting. Long-term care providers and information systems vendors focus their systems development and implementation efforts on meeting government requirements. The resulting systems usually offer little support for clinical documentation and order management. The long-term care census component is not updated on a real-time basis, since patient management (admitting, discharge, and transfer) data are often entered just before the billing process. Demographic information in the finance system is usually very basic. Financial management varies by facility. Payroll, for example, could be managed by a service firm with in-house general ledger, accounts payable, and accounts receivable on a microcomputer.

If the facility is part of a chain, all financial processing may be done in corporate headquarters. Network links between acute care and long-term care are in their infancy.

Until recently it was not really cost-effective to extensively use computers to manage long-term care information.[22] Care patterns and recordkeeping requirements were not as complex as in acute care or specialty hospitals. Resident populations were more stable and generated a smaller volume of information. Lack of financial resources placed long-term care behind the rest of health care in terms of automation. The high start-up costs for hardware, software, implementation support, and training as well as the ongoing cost of systems maintenance put comprehensive information systems solutions out of reach for the traditional long-term care facility.

Continuous operational changes make it difficult to realize the full potential of automation in the long-term care setting. Restructuring to comply with government regulations is constant. High employee turnover makes it difficult and unprofitable to train long-term care workers to use computers. Employees sometimes resist systems because of the resulting changes in familiar practices and procedures. The assignment of systems administration and report-writing responsibilities can pose problems. Furthermore, the time-consuming process of selection and implementation of a system adds strain to long-term care facilities that may already be short-staffed or lack experienced information systems personnel.

The broad array of available long-term care information systems products only complicates improvement efforts. At least 60 long-term care vendor product offerings are on the U.S. market today. State-specific long-term care regulations result in the proliferation of vendors, most catering to a small group of clients. Regionalized information systems vendors offer products ranging in capabilities from minimum payment classification and quality monitoring to comprehensive resident care systems with electronic resident charts.[23] The steady stream of revisions in regulatory requirements consume vendor design and program resources and necessitate rigorous client support. Development of systems with equally strong clinical and financial functionality and interfacing to contracted services presents yet more challenges. Generally, such promising applications are in the early stages of the product life cycle and will eventually become more uniform in design, features, and price.[24] In the meantime, vendor selection is a daunting prospect for the long-term care provider.

Steps ahead. Despite all the formidable barriers to information systems design and implementation, most long-term care facilities today have some degree of automation. But utilization varies widely among installed systems.

Some use systems to meet regulatory requirements for electronic submission of assessment data. Certain states provided free software for this purpose, but reporting functions are limited. Other facilities have purchased systems that link MDS/MDS+ and care planning using the mandated RAI. Assessment data are processed to provide the appropriate RAP. For long-term care facilities that already have well-defined resident assessments and comprehensive care plans, regulations are incorporated into the existing tools. The advantages of this approach include user-defined screens, pathways, and report formats. However, such systems call for increased setup and maintenance responsibilities for the user.

Finally, some long-term care facilities are moving ahead into integrated assessment, care planning, order scheduling, and progress note automated systems intended to replace paper. These users are setting the pace for the rest of the industry.

Evaluation and Implementation of Automated Solutions

With a clear understanding of the current state of long-term care and prevailing industry trends, development and implementation of information systems solutions can begin. This section offers a practical approach to developing and implementing information systems solutions for long-term care. It includes long-term care information systems possibilities, guidelines for an operations analysis, a discussion of information systems approach issues, information systems product and services evaluation criteria, and an overview of system implementation in long-term care facilities.

Evaluation and implementation time frames, budgets, and priorities will, of course, vary for every institution. Our intent is to present a logical progression of activities that lead to the successful evaluation and implementation of information solutions in long-term care.

The Information Vision

Clarifying a vision is the first step toward an advanced long-term care information systems solution. Planners should conceptualize the latest information technology to achieve entirely new long-term care goals. Such a strong vision will serve as a constant reference point throughout the process of selecting and implementing a system.

Envision an information management system that provides the flexibility to adapt to changing market demands and the sophistication to fully integrate resident, demographic, clinical, and financial information.[25] At the long-term care facility level, such a comprehensive information management system requires universal interface/network architecture to receive, validate, reformat, and send data between department systems and end-user universal workstations. Information could then be entered just once and used in many different ways.

Streamlining the flow of vital information. A computerized medical record for a resident could include assessments, care plans, progress notes, medications, case mix reimbursement analysis, accident-incident monitoring, and infection control information.[26] Aggregate storage of this data in an electronic repository would allow access and sharing of information from any location. A data repository would also free the department systems from the burden of data storage, improving individual workstation performance.

At the network level, a Delivery-Wide Health Information Network linking physically independent healthcare entities allows appropriate data and information flow, thus making information available anywhere at any time. The long-term care entity then benefits from data captured at physician offices, independent pharmacies or labs, home health agencies and acute care facilities.

Operational improvements. From an operational standpoint, long-term care consumers, clinicians, administrators, and payers could expect lower costs and greater efficiency with improved long-term care information management. The long-term care consumer would enjoy improved health status, reduction in unnecessary medical care, and better coordination of services when treatment is required. Access to information about a variety of healthcare options would aid long-term care consumers in the selection of providers and services. Automated data collection and analysis could also facilitate improved outcomes research as well as medical advances that improve resident care. Finally, interactive consumer health information could educate long-term care consumers and promote wellness.

Clinical benefits. For long-term care clinicians, easy access to clinical information and treatment protocols would provide improved clinical decision support. Better decision support in the form of system-generated alerts or warnings, for example, can minimize redundant diagnostic testing and reduce adverse drug reactions. Over time, this type of dynamic feedback would reinforce resident care standards. Communication among members of the care team would be enhanced by automated monitoring, evaluation, and management of treatment practices and care plans. In addition, on-line access to aggregate resident information would enable clinicians to easily customize treatment protocols. Subsequent updates, benchmarks, and trend analyses would be easier.

Computerization could also offer information processing methods that promote clinician efficiency.[27] For example, MDS/MDS+ work is ideally suited for computers. Resident assessment data are collected and processed through a large decision tree. Certain combinations of answers, when sifted through the decision tree, may show that the resident has a problem listed in the RAP summary. If the resident has one of the problems on the summary, treatment will be reflected in the resident's care plan. A process that now takes two hours to complete manually could be done in 15 to 30 minutes on a computer.[28] The system could even report when the full MDS/MDS+ was complete and provide due dates for quarterly and annual reviews.

Administrative and financial benefits. For long-term care administrators and third-party payers, information technology can facilitate quality and cost control, streamline the referral process, and improve cost management capabilities via easy access to benefits eligibility information. Preadmission screening and annual resident review program could be implemented to help determine appropriate placement.[29] Automated MDS/MDS+ could serve as the foundation for acuity-based staffing, quality assurance monitoring, and research.

Long-term care administrators and third-party payers would have better outcomes measurement, reporting, and management capabilities. Electronic claims submission and funds transfer would improve cash flow management. Department managers would have access to pricing systems that support estimation of resource needs, charges, and associated costs. Electronic linkage to knowledge databases can increase the ability to compare best practices and enhance tracking of physician practice patterns. Automated critical pathways could provide the basis for measuring compliance with organizational, state, and federal quality standards. Accurate and timely MDS data collection could maximize reimbursement potential and help simplify case management and utilization review.

An automated information systems network has the potential to reduce paperwork, minimize duplication, and increase processing speed. Administrative efficiency would improve by using computer-generated forms and electronic communication. Other administrative benefits could include improved staff and shared resource scheduling, better contract management, and enhanced strategic decision support.

Analyzing Operations

With a future vision firmly in mind, planners may proceed to analyze current long-term care business practices to develop information systems solutions. The operations analysis is conducted in five steps:

1. *Establish a project steering committee.* The project committee will provide executive sponsorship, overall project direction, and final approval for the solution.

2. *Assess current strategies.* Begin the analysis by assessing the current strategic plan, including mission, financial and service plans, building and technology plans, long-range vision, and nursing care strategies at the single-facility and network levels.[30] Evaluate the knowledge, experience, and track record of the current leadership team. Review organizational charts. Information solutions must be selected and implemented within the proper strategic context.

3. *Identify all major work processes and flows, policies and procedures, and resources and tools,* such as forms, reports, and computers, used to support the work processes as well as associated revenues and expenses. Structure the analysis so that all functional areas of the organization are reviewed. Include representatives from each area touched by the work processes. Consider long-term care processes that pertain to referrals, admissions, precertification, census management, clinical assessment and care planning, claims processing, resource scheduling, and statistical reporting. The information systems component of the review should include the present level of automation, a complete systems portfolio, and a comparison with industry standards and state requirements. Consider intraindustry dynamics as well as the forces driving change in the use of automation in the long-term care facility, such as managed care, government regulations, case mix reimbursement, and network affiliations. The outcome of this phase of the operations

analysis will include a detailed written description of processes and work flows along with supporting documentation, such as forms, policies and procedures, current staffing levels, and statistics and financial reports.

4. *Develop clinical, financial, and administrative performance and quality improvement objectives* based on analysis findings. It is from these objectives that the information systems solution must evolve and by which the organization's return on its information technology investment must be measured. Once the performance objectives have been approved, study the objectives to uncover information systems priorities. Develop a list of explicit needs for a system (requirements definition) to aid in the preparation of information systems solution options.

5. *Obtain executive approval* to evaluate information systems solution options.

An objective operations analysis identifies facility strengths and improvement potential. It is used to describe the effectiveness of the organization and to lay the groundwork for improved information management. It demonstrates the need for change. Moreover, it serves as a benchmark to accurately assess the value of the information systems solution once it is implemented.

Weighing Options

The future vision, operations analysis, performance improvement objectives, and systems requirements definition provide a framework for selecting a viable information systems solution. Setting information systems parameters will help define the scope of acceptable options. Performing a vendor and product analysis will further narrow the field of solution alternatives.

Setting system parameters. First, address the major information systems decisions and set the parameters for evaluating options:

- *Operating platforms: mainframe, minicomputer, or PC-based distributed networks?* Distributed systems offer more sophistication at the end-user level. However, cost management is a key element of this decision. For mainframe systems, 80 percent of the lifetime cost is consumed in a one-time hardware investment; software and maintenance account for the remaining 20 percent. With distributed systems, it is just the opposite.[31] With the mainframe option, remote or on-site location must also be considered.

- *Single-vendor or multiple stand-alone systems?* Although there are feature advantages to selecting a "best of breed" product, system integration and lack of cooperation among vendors can cause major problems. Single-vendor solutions offer facilitywide connectivity but may not provide application depth or department-specific vendor solutions. Single-vendor solutions also require an appreciation of how global system changes affect each user area.
- *Centralized or decentralized approach to automation?* Generally, the decentralized approach puts more control at the user level. Users have greater access to information when they need it, but they also have a higher level of accountability.[32] The advantage of a centralized approach is that it frees users from system responsibilities.
- *Invest in facility-specific solutions or form alliances to obtain automation?* Forming alliances could involve vertical integration with an acute care facility or becoming part of a multifacility chain. The long-term care entity may trade access to greater resources for system flexibility.
- *Separate or integrated financial and clinical information systems?* Clinical systems requirements are subject to interpretation by clinicians. They are facility-specific and are always changing. For these reasons, there is a tendency to implement a separate resident care system. But since Medicaid and Medicare reimbursements rely on MDS+, an integrated approach may be the better choice.[33]
- *A staged approach versus the "big bang" approach to implementation?* Available time and resources as well as vendor capabilities should be factored into this decision.
- *Provisions for maintaining confidentiality?* Providing ready access to resident information has to be balanced with the responsibility of protecting residents' privacy. Facility policies and procedures, current laws, and system capabilities must be considered.

Selecting a vendor. Next, evaluate long-term care information systems vendors and their products and services. The company that develops, markets, and supports the long-term care software is just as important as the product. Number of years in business, corporate philosophy, profitability, number of installed clients, market share of like institutions, dedication to the long-term care industry, knowledge of government regulatory requirements (MDS/MDS+), and staff expertise are factors to consider in the selection of a long-term care vendor.

The product evaluation process is a major component of solution evaluation. There are several ways to approach this phase. Systems buyers often provide vendors with a request for information (RFI) that includes automation objectives, facility overview, information systems requirements, and response instructions. Vendors complete the RFI, indicating to what extent their products and services will meet facility needs. Buyers compare vendors, based on their RFI response, to narrow the list of vendor candidates. A vendor may be eliminated for any number of reasons—bad references, lack of features in the software, or higher price than offered by other vendors.

After the short list of potential vendors has been established, system demonstrations are scheduled. Demonstrations provide the opportunity to learn more of the system's strengths and weaknesses relative to facility requirements. To enhance the value of this phase, system requirements should be further developed into either detailed checklists or realistic scenarios. When demonstrations have been completed, buyers visit installed sites to evaluate the vendor product and interview current customers in a "live" setting.

Table 7-1 presents a detailed list of criteria for selecting long-term care products and services.

Testing for feasibility. Evaluating the information systems solution's feasibility is the final phase of options analysis and system selection. Operational feasibility indicates how well a specific solution will work in the organization. Technical feasibility measures the practicality of a specific technical solution, the availability of technical resources, and the feasibility of the implementation schedule, if deadlines have been established. Finally, economic feasibility demonstrates the solution's cost-effectiveness via a cost-benefit analysis.

TABLE 7-1 Selection Criteria for Long-Term Care Information Systems Products and Services

Software

— Complete list of software applications and modules
— Uses proven turnkey systems
— Percent of in-house development
— Meets industry standards
— Ease of use
— Response times
— Depth of library of care plans
— Flexibility (Can it be customized?)
— Access to information through report writers and query tools
— Generate reports and queries without a programmer
— Data integrity, security, and confidentiality
— Output formats ("look and feel")
— Interface and remote access capabilities (specifications clearly stated)
— Database management
  • Size and flexibility
  • Ability to modify
— Scheduled downtime, backup, and data recovery
— Data storage and archiving (note state requirements for on-line MDS/MDS+ record access)
— Source code available in case of bankruptcy
— Third-party software

Hardware/equipment

— Compatible with existing systems and hardware
— Industry standard file formats
— Modem required for transmission to state or on-line vendor support
— Accommodate state-required media
— Adequate storage
— Printer type—able to print required forms
— Protection
  • UPS
  • Surge protector
— Site preparation
— Hardware maintenance
— Warranties, titles, and modifications
— Equipment failure—field engineer on site within specified hours of notification
— Loaned equipment
— Operating systems

Installation/implementation services

— Extent of implementation support
— Implementation schedule/work plan
— Ability to evaluate (and reject) vendor installation team members
— Effective date
— Adherence to deadlines
— Proper sign-offs

(continued)

TABLE 7-1 Selection Criteria for Long-Term Care Information Systems Products and Services *(continued)*

Training

— Methodology—"train the trainer"
— Location
— Duration
— Materials required
— Experienced instructors
— Current user manuals delivered to facility
  • Number of copies needed
  • Policy for periodic updates
— Training support for new employees after implementation

Ongoing customer support

— On-line support (24 hours, 7 days a week)
— No charge for regulatory updates
— Enhancements and upgrades
— Response time and escalation process
— Service and support track record
— Remote diagnostics
— Annual support renewal
— Contract termination
— Backup tape verification
— Communication via newsletter or other means to user groups
— Conversion assistance in the event of termination

Customer referrals/site visits

— Similar institution
— Number of months or years installed
— Level of utilization
— Level of satisfaction
— Vendor support in problem resolution
— Degree of customization
— System reliability

Price/payment schedule

— One-time start-up costs
— Ongoing costs: software, hardware, implementation, training, support
— Payment linked to acceptance
— Payment schedule based on work plan
— Cap for allowable price increases
— Payment penalties for nonperformance
— Daily expenses (travel, food, etc.)

Other costs (direct and indirect)

— Site preparation (cabling, wiring, outlets, phone lines)
— Consultants
— Office furniture (cabinets, chairs, printer stands)

(continued)

TABLE 7-1 Selection Criteria for Long-Term Care Information Systems Products and Services *(concluded)*

— Telephones, headphones
— Office supplies (paper, printer cartridges, etc.)
— Temporary staff to cover for regular staff during system training
— Retrospective data entry
— Customizations
— Initial inefficiencies
— Human resistance slowing the process

Excluded services

— Electric work external to equipment
— Repair of equipment resulting from accident or negligence
— Services associated with relocation
— Routine cleaning
— Repair or replacement of defective magnetic diskettes, tapes, and ribbons used with equipment
— Major overhauling of system needed after 5 years
— Services performed outside of business hours

The prospects for reimbursement are best examined at this point. Table 7-2 presents items to consider in seeking Medicaid reimbursement for assistance in selecting and implementing long-term care management information solutions.

Quantify the feasibility study as much as possible. Compare strategic goals; operational objectives; information systems priorities; risk assessment; and resource, budget, and time constraints for each solution option.

Documentation and approval. Planners should prepare a written analysis of every option they consider. Include option parameter decisions, vendor finalists, and written feasibility analyses along with the ultimate recommendation for option selection. Explain the rationale for the recommendation, including estimated time frames, estimated capital and budget expenditures, and key assumptions. Itemize facility benefits and risks. Describe how performance will be measured and success evaluated. Include staffing requirements.

Finally, information systems planners may obtain executive approval to complete the options analysis and negotiate a contract with the vendor of choice.

TABLE 7–2 Medicaid Reimbursement for Assistance in Selecting and Implementing Long-Term Care Management Information Solutions

Medicaid reimbursement options vary from state to state. Any of the following stipulations may apply in your state.

- All assets, including information systems equipment and software, are depreciated over their useful lives on a straight-line basis and in accordance with American Hospital Association guidelines.
- Useful lives may be adjusted if the provider submits a written request.
- Long-term care providers enjoy no opportunity to have depreciation related to any capital addition (except building assets) reflected in their current Medicaid rate of payment.
- Capital is a pass-through expense. The state will reimburse eventually through retrospective settlement. Check on whether limits apply and on timing.
- Providers possibly may be reimbursed for depreciation and interest expense related to any capital addition reflected in the current Medicaid rate of payment, if the capital asset addition is required by the government.
- Providers may possibly be reimbursed for depreciation and interest expense if the state is given proper notification before purchase and if bed thresholds are met.
- Providers in some states may request an increase in Medicaid's interim rate of payment to cover depreciation and interest expenses related to a long-term care information system. Submit a written request with documentation indicating the additional expense being incurred.
- In some states, providers must wait until the base year is updated to have additional depreciation reflected in their Medicaid rates of payment. Base years are not typically updated on a regular basis; it depends on state law. Expenses may be reflected in the rate when the acquisition year becomes the base year.
- Interest expense related to movable equipment is not reimbursable. For-profit facilities may be eligible to receive a return on equity. Not-for-profit facilities do not receive a return on equity in some states.
- Reasonable consultation assistance expenses related to the selection and implementation of information systems generally are allowable. Such expenses are considered part of the asset and are capitalized in accordance with generally accepted accounting principles.

Implementation

Now the real work begins! Implement the most feasible long-term care information systems solution. Fortunately, much of the implementation planning tasks have already been accomplished during the options selection process. Implementation services, a detailed work plan, milestones and time frames, and payment schedules are developed as part of the contract negotiation. The experience of the implementation team, competent vendor support, executive sponsorship, and end-user involvement then become critical implementation success factors. While the complexity and duration of tasks may vary, system implementation follows a predictable course.

Primary implementation activities. Because project management tasks take place throughout implementation, dedicated project managers work at the long-term care facility and on the vendor side. Their responsibilities include leading the team, monitoring progress, and revising the plan as well as communicating project status to the executive steering committee.

Site preparation. Regardless of the solution options selected, some type of site preparation will be required. The work flow analysis and hardware requirements will help determine equipment location, cabling network specifications, and electrical requirements. Remember to factor in space and wiring for print devices as well as the workstation.

Delivery, installation, and validation. Delivery, installation, and valid-ation of hardware and software are project milestones. Timing of these events will depend on how information is currently being handled, available storage space, and security arrangements.

System specification and table/file building. Once hardware and software have been installed and validated, system specification and table/file building can begin. Standard table setup, screen enhancements, custom reports, and vendor responsibilities should be discussed during solution option evaluation and contract negotiation. If not closely managed, this phase of the project could consume many more hours than planned. Keep track of all system changes requested. Identify changes that provide the most benefits and must be completed; then make those changes first. Work with the vendor to establish an organized approach to this phase.

Interface specification and development. Interface specification and development can be performed concurrently with system specification. Review the rationale for the interface, standard versus customized options, and the vendor's track record in this area.

System testing. Another critical component of the implementation plan is system testing. Deciding if and when to convert existing records and resident care plans will depend on the system testing strategy, available resources, and vendor product offerings. Electronic file transfers are timesaving but not foolproof. If manual data entry is required, investigate the possibility of hiring temporary assistance.

Employee training. Plan employee training to occur shortly before systems conversion so that staff members can exercise their newly learned skills. Develop or modify training tools to fit facility needs and provide supplemental resources to cover work areas during training.

Policies and procedure review. Policies and procedures will have to be revised or created to incorporate the new system into the long-term care facility. Factor in sign-off requirements and continuous policy and procedure review into the implementation plan.

Conversion. The actual conversion or migration from the old system to the new system is always an exciting event. Carefully plan through the details of this process. Consider extra staffing, "go/no-go" checkpoints, and fallback plans.

Postimplementation review. Often easy to forget, the post-implementation review is an important component of the implementation plan. Performing a post-"live" analysis provides the information needed to correct system problems, refine functionality, and measure the system's impact on operations.

Table 7–3 presents a checklist of the implementation tasks discussed in this section.

Keys to Success

Let there be no doubt: the evaluation and implementation of a long-term care information solution is a complex, time-consuming, and expensive process. Executing the implementation plan on schedule and within budget, deploying the technology, acknowledging the nontechnical aspects of system implementation, and evaluating the system's impact on operations will determine the success of the information solution in daily practice.

TABLE 7-3 System Implementation Checklist

Implementation project management (over project duration)

— Perform implementation planning during contract negotiation
— Identify implementation goals
— Identify implementation tasks and time frames
— Assemble an implementation project team
  • Create an executive steering committee or designate an executive sponsor
  • Assign a dedicated long-term care information systems project manager
  • Identify an information systems project manager on the vendor side
  • Specify resource commitment requirements on both sides
  • Create an organizational chart for the implementation team
  • Specify skills, roles, relationships, and team structure (user representatives, system analysts, system administrator, etc.)
— Identify a process to guide the implementation project
  • Confirm implementation tasks, responsibilities, and work effort via the work plan
  • Determine the order in which applications will be installed and implemented
— Monitor the progress of the work plan and report back to management
  • Develop a project communication strategy
  • Track actual hours and compare with planned work efforts
  • Manage expectations regarding system availability and functionality
  • Create a mechanism for issue identification, escalation, and resolution
  • Create a mechanism to manage system modification requests
  • Update the work plan as necessary

Site preparation

— Specify hardware requirements during contract negotiation
— Assemble appropriate resources to develop a site preparation plan
— Plan equipment layout at every user site
— Perform facility upgrades (wiring, ventilation, temperature control, etc.)
— Test site alterations
— Coordinate site preparation with hardware delivery and installation.

Hardware delivery and installation

— Set delivery dates during contract negotiation
— Confer with the vendor regarding hardware delivery and testing plans
— Plan for hardware storage before installation
— Test hardware to ensure functionality
— Document results and notify vendor

Software installation and validation

— Set delivery dates and criteria for acceptance during contract negotiation
— Develop a plan for access to software for validation, system development, and training
— Define system backup and maintenance procedures
— Confer with the vendor regarding the software validation test plan and test materials
— Perform the software validation test
— Document results and notify vendor

(continued)

TABLE 7-3 System Implementation Checklist *(continued)*

System specification

— Review required business processes from the analysis phase of system selection
— Identify current information management support (manual and automated) for these processes
— Compare current information management to new system capabilities
— Map support needed to new system equivalents
— Identify processes and information requirements not supported by the new system
— Identify customization required for implementation
— Identify level of customization required
  • Work effort
  • Cost
  • Resources responsible for customization
  • Impact on critical path
— Determine which customizations will be available for system implementation
— Create a postimplementation customization list
— Continue to manage user expectations

Master file/table building

— Establish level of vendor support during contract negotiation
— Determine work effort and timing during contract negotiation
— Meet with technical and user representatives to determine profile values for system master files
— Document profile values
— Monitor table construction
— Ensure table accuracy and completion
— Develop table maintenance procedures

Interfaces

— Clearly define interfaces during contract negotiation
  • Purpose
  • Systems affected
  • Features and functional capabilities
  • Development costs
  • Installation costs
  • Testing costs
  • Past experience with specific different vendors and interfaces
— Differentiate between department interfaces and interfaces to contracted services (e.g., laboratory and pharmacy) and to acute care providers
— Develop interface specifications
— Build interfaces
— Test interfaces

Conversion of existing records, resident care plans, and orders

— Establish workload, time required, and vendor support during contract negotiation
— Develop plan for uploading existing records, care plans, and orders
— Identify resources to perform data entry

(continued)

TABLE 7-3 System Implementation Checklist *(continued)*

— Perform data entry
— Determine disposition of existing documents once entered
— Print audit reports to verify accurate data entry

Testing: unit, integration, interface, and acceptance

— Establish testing schedule during contract negotiations
— Confer with vendor regarding
  • Test plan development and expected outcomes
  • Testing tools (checklists, scenarios, etc.)
  • Test result reporting and problem resolution
  • Vendor support and availability during testing
  • On-line training/testing "area" that will not affect live records
  • Payment associated with testing milestones
— Perform testing activities
— Evaluate test results
— Resolve testing issues
— Ensure system's functional and technical acceptance

Review of policies, procedures, and operations

— Review current work flow, organizational structure, and policies and procedures from the analysis phase of system selection (include facilitywide as well as department-specific policies and procedures)
— Identify changes to procedures based on system constraints
— Make revisions to incorporate new system into normal business operations
— Create new policies and procedures when necessary
— Obtain sign-off on revisions and additions
— Develop a plan for continuous policy and procedure review and updates
— Introduce revised or new policies and procedures
— Make revised or new policies and procedures accessible to users

Employee training

— Establish training approach and vendor support during contract negotiations
— Assign training coordinator
— Identify "super users"
— Confer with vendor regarding
  • Training location
  • Time requirements
  • Training materials (scenarios, checklists, evaluation tools, etc.)
— Develop a training plan (implementation and ongoing training support)
  • Assess current level of knowledge
  • Establish learning objectives
  • Develop a training methodology
    — Initial learning and validation
    — Department-specific training (e.g., claims processing)
    — Function-specific training (e.g., report writing)
  • Ensure system availability
— Scheduling training sessions
— Perform employee training
— Update training methods to comply with system changes

(continued)

TABLE 7-3 System Implementation Checklist *(concluded)*

Conversion strategy
— Establish vendor support during contract negotiation
— Prepare a plan to convert from existing systems (manual or automated)
  • Migration approach (i.e., parallel or complete conversion)
  • Time required
  • Schedule to cause minimal disruption of services
  • Extra resources required
  • Employee preparation
  • Alternate, or fallback, plan
— Execute conversion plan: Go live!

Postimplementation review
— Verify vendor support during contract negotiation
— Confer immediately after implementation for a progress report
— Identify action steps to smooth implementation
— Celebrate!
— Reconvene 3–6 months after implementation; analyze system's impact
  • Determine whether the system is meeting organizational objectives
    — Perform interviews with users
    — Review reports
    — Observe work flow
  • Assess users' level of satisfaction
  • Assess level of system utilization
  • Identify improvement opportunities

Beyond the guidelines prescribed above and in Table 7-3, operations systems planners would do well to heed the following words of advice for smooth implementation.

Manage expectations at the executive and user levels. Improved efficiency and cost savings through information systems assumes maximum efficiency of the system and complete adoption by staff. In reality, change happens slowly for the end user. Executives should expect savings to be reduced during the early years by the costs of acquisition and training. Involving end users throughout selection and implementation and managing the implementation could shorten the adoption schedule and thereby substantially increase the eventual savings from the new information technologies.[34] Reinforce the idea of greater value over time.

Users need to be prepared for the frustration that so often accompanies the excitement of new information management systems. This is especially true if system implementation requires users to make a transition from manual to automated procedures. System utilization and user acceptance will be enhanced by honestly discussing system limitations, developing acceptable "work-around" options, formally identifying and resolving issues, ensuring comprehensive training, and incorporating user input in the migration strategy.

Understand the limits of automation. Information technology surmounts the barriers of time and distance. It enables organizations to break old rules and create new ways of thinking.[35] However, selection and implementation of the right information systems technology are just parts of the total solution. Ongoing executive sponsorship, a commitment to improving resident care, and a willingness to embrace change are critical factors toward success. Integration has to happen at the relationship level first. In the end, it is the system users, thinking through the problems and using technology to expedite solutions, who determine the lasting results of any information systems solution.

Recognize that some things will always change. Government regulations, resident acuity, population demographics, and documentation standards are just a few of the forces that are constantly reshaping long-term care. As the nature of the problems change, so will the nature of the solutions. The cycle of current-state analysis, solution evaluation, and implementation is perennial. Retooling the long-term care facility with flexible information systems should position providers to respond intelligently to all the changes that lie ahead.

Recognize that some things will always stay the same. In long-term care, the resident always comes first. Ultimately, information systems solutions in long-term care are designed to facilitate care provided over an extended period to people, regardless of age, with chronic disabilities. Responsive long-term care solutions should therefore help care providers enhance residents' quality of life and promote autonomy.

Summary

Our challenge is to create a systematic, predictable, and organized approach to delivery of service that is also flexible and responsive.[36] The goal is to increase productivity by coordinating departments, programs, and entities across a continuum of care to lower costs and improve quality of service. The end result will be a customer-responsive healthcare delivery system supported by integrated information management.

References

1. Anders, K. T. "Computerized Caring," *Contemporary Long-Term Care*, August 1994, p. 8.
2. King, R. T., Jr. "The Business Plan: The Magic Formula," *The Wall Street Journal*, November 14, 1994, p. B1.
3. Hammer, M. and J. Champy. In *Reengineering the Corporation: A Manifesto for Business Revolution*. New York: Harper Collins, 1993, pp. 83–101.
4. Charlson, J. T. "Establishing Links with Long-Term Care Facilities," presented at the American Hospital Association's 1989 Healthcare Information and Management Systems Society Conference.
5. Charlson. "Establishing Links."
6. *Introduction to Long-Term Care: An Independent Study Program*. Cleveland: Ernst & Young, May 1993.
7. *Introduction to Long-Term Care*.
8. *Introduction to Long-Term Care*.
9. *Introduction to Long-Term Care*.
10. Charlson. "Establishing Links."
11. Roland, D. L. "Clinical Computerization in Long-Term Care Facilities." In *Long-Term Care Administration Handbook*. Gaithersburg, Maryland: Aspen Publishers, 1993.
12. *Introduction to Long-Term Care*.
13. *Introduction to Long-Term Care*.
14. *Introduction to Long-Term Care*.
15. *Introduction to Long-Term Care*.
16. *Introduction to Long-Term Care*.
17. Roland. "Clinical Computerization."
18. Gardner, E. "Nursing Homes Get Pushed Closer to the Computer Age." *Modern Healthcare's Eldercare Business*, November 1990, p, 10.

19. Gardner. "Nursing Homes," p. 00.
20. Roland. "Clinical Computerization."
21. Roland, "Clinical Computerization."
22. Gardner. "Nursing Homes."
23. Gardner. "Nursing Homes."
24. Hume, S. K. "Integrating Services for the Elderly." *Health Progress,* October 1993, p. 8.
25. Roland. "Clinical Computerization."
26. Anders. "Computerized Caring."
27. Roland. "Clinical Computerization."
28. Gardner. "Nursing Homes."
29. Hegland, A. "Computers Empower Nurses." *Contemporary Long-Term Care,* August 1993, p. 8.
30. Simpson, R. L. "Why an I.S. Business Plan?" *Nursing Management,* March 1993, p. 3.
31. King. "The Business Plan."
32. Roland. "Clinical Computerization."
33. Hume. "Integrating Services."
34. Henrion, M. and J. Silva. "Cost Savings from Information Technology in U.S. Healthcare Reform: Insights from Modeling," *Journal of the Healthcare Information and Management Systems Society,* Winter 1994, p. 1.
35. Hammer and Champy. *Reengineering the Corporation,* p. 87.
36. Hume. "Integrating Services."

For Further Reading

Durand, D. "The Short Course: Long-Term Care Automation," *Healthcare Informatics,* September 1994, p. 9.

Guide to Effective Healthcare Information and Management Systems. Chicago: Healthcare Information and Management Systems Society, 1994.

Acknowledgment

The author wishes to acknowledge Tom Haslett, Susan Abba, Christina Braman, and Bob Heatley for supporting her research and writing here.

CHAPTER 8

Demonstrating Financial Feasibility

Roy W. Byers

Every long-term care development, whether the construction of a totally new facility or the expansion of an existing campus, involves a wide array of planning steps and analyses that require the expertise of a variety of healthcare professionals. The lengthy development process evolves in stages, from product and service conceptualization through land acquisition, building design, construction, and fill-up.

The sponsors or developers of a long-term care community accept financial responsibility for its future success, a responsibility that extends to the community's residents, staff, and financiers. Even with the best efforts of able professionals, the sponsors or developers cannot fulfill these responsibilities or realize their vision of a long-term care community without first demonstrating that the project will be financially feasible.

The financial viability of the proposed community must be demonstrated to the satisfaction of all interested parties, including the management team, directors, regulators, potential residents and their advisors, the financing team, and financiers. Identifying key assumptions and testing those assumptions for reasonableness are integral parts of the feasibility process. Depending on management's experience and sophistication and on the project's stage of development, the planning team may find it necessary to engage a feasibility consultant to help formulate assumptions regarding the long-term community's future operating and financial performance—or, at a minimum, to test the reasonableness of such assumptions.

This chapter presents a step-by-step guide to demonstrating the financial feasibility of a long-term care development. It explores the purposes and objectives of the feasibility study; accepted methods for conducting a study and reporting results; and the roles of management, consultants, and other professionals.

Purpose of the Financial Feasibility Study

The financial feasibility study is designed to provide interested parties with relevant market and financial information on the potential performance of a new or existing long-term care or retirement housing facility. The interested parties and their use of the financial feasibility study will differ depending on the developmental stage of a project. In general, the management and the trustees or directors of an organization constitute the interested parties who initiate and have ultimate responsibility for the financial feasibility study. These interested parties may be motivated by either an internal or an external need to demonstrate financial feasibility.

Table 8–1 summarizes some of common purposes for feasibility studies and the appropriate level of study for each. The differences between the types of reports are discussed in the sections "Feasibility Study Reports" and "The Financial Feasibility Consultant's Services."

Prospective Financial Information

Simply put, prospective financial information describes the future. Prospective financial information may be classified by its form of presentation or by the nature of its description of future states.

Prospective financial information may include a complete presentation of all basic financial statements and footnotes or a partial presentation of certain financial information, financial statements, or

TABLE 8–1 Purposes and Types of Feasibility Studies

Purpose	Type
Management planning, market and financial planning	Preliminary market analysis
	Preliminary market analysis
Board review and approval	Full feasibility study, compilation, or examination
Regulatory review with external consultant	Full feasibility study, compilation, or examination
Marketing with outside consultant	
Financing with outside consultant	Examination

footnotes. Generally, a study intended for external distribution requires a full presentation of financial statements and footnotes, while partial presentations often suffice for strictly internal purposes.

Prospective financial information may represent either a projection or a forecast. Projections are prospective financial statements that present, to the best of the responsible party's knowledge and belief, an entity's expected financial position, results of operations, and cash flows, given one or more hypothetical assumptions. Hypothetical assumptions are not necessarily the most likely to occur, and management should take pains not to misrepresent the supportability of its assumptions. The financing rate for a project in the early stages of analysis, before a form of financing is resolved and a financier is identified, is one example of a hypothetical assumption.

Forecasts are prospective financial statements that present, to the best of the responsible party's knowledge and belief, an entity's expected financial position, results of operations, and cash flows, based on the responsible party's assumptions regarding expected future conditions and courses of action.

Management may use financial projections in the planning stages of a project and later, as assumptions are further clarified and become supportable, convert the projections into a financial forecast. In most cases, outside users of prospective financial information will require financial forecasts.

Feasibility Study Reports

The feasibility study consultant may issue a variety of reports expressing various degrees of analytical rigor and certainty in the study results, assumptions, and consultant's opinions. In its *Guide for Prospective Financial Information,* the American Institute of Certified Public Accountants (AICPA) establishes standard policies, procedures, and reporting requirements for feasibility study consultants. The guide provides for four levels of reporting:

1. *Examination.* The consultant expresses an opinion on the reasonableness of the assumptions and on the presentation. This is the highest level of reporting and provides for a separate form of report letter for a feasibility study.

2. *Compilation.* The consultant indicates the compilation service provided but disclaims opinion or giving any form of assurance on the reasonableness of the assumptions and financial statements.
3. *Agreed-upon procedures.* The consultant indicates the procedures performed with respect to the prospective financial statements and may provide negative assurance with respect to the specific procedures. The consultant disclaims opinion or giving any form of assurance on the reasonableness of the assumptions and financial statements.
4. *Assembly.* The consultant indicates the assembly service, disclaims compilation and examination service, and provides no assurance on the reasonableness of the assumptions or the financial statements.

The appropriate level of reporting will partly depend on the needs of the client and the audience and intended use of the report. A report may be issued for general use, limited use, or internal use only.

A general-use report is any report that may be used by people with whom management is not directly negotiating. Ordinarily, recipients of a general-use report would not have the opportunity to negotiate directly or ask questions of management. General-use reports are commonly used for public offerings of debt or equity.

Limited-use reports are prepared for use by third parties with whom management is negotiating directly. These individuals should be knowledgeable about the industry and should have the opportunity to ask management direct questions. Often, third-party, limited-use reports are used to negotiate the terms of a financing.

Feasibility consultants frequently are engaged to prepare prospective financial information intended solely for the internal use of management. Internal-use-only reports may be used, for example, in evaluating the financial viability of significantly expanding a retirement community or adding a new level of care.

Depending on the type of feasibility report prepared and issued, the use of forecasts and projections within the context of a financial statement or a partial presentation may be appropriate or inappropriate. Table 8–2 suggests instances when each report is most appropriate.

TABLE 8-2 Appropriate Use of Feasibility Study Report Types

		Financial Statement		Partial Presentation	
Use	Report	Forecast	Projection	Forecast	Projection
General	Examination	Yes	No	No	No
	Compilation	Yes*	No	No	No
	Agreed-upon	No	No	No	No
	Assembly	No	No	No	No
Limited	Examination	Yes	Yes	Yes	Yes
	Compilation	Yes	Yes*	Yes*	Yes*
	Agreed-upon	Yes*	Yes*	Yes*	Yes*
	Assembly	No	No	No	No
Internal	Examination	Yes	Yes	Yes	Yes
	Compilation	Yes	Yes	Yes	Yes
	Agreed-upon	Yes	Yes	Yes	Yes
	Assembly	Yes	Yes	Yes	Yes

*Appropriate in certain circumstances.

Feasibility Study Consultant's Services

The feasibility study consultant can provide a wide range of services in the development of a long-term care or retirement housing facility. The types of services that the consultant provides will depend on the client's needs and the developmental stage of the community.

The initial phase of a feasibility consultant's work with a retirement community may entail a preliminary market analysis. During this phase the consultant evaluates the potential need in the market for the types of retirement housing services that a sponsor intends to offer. In a preliminary market analysis, this need is tested on a purely numerical basis and does not usually involve substantial interviews.

If the preliminary the market analysis indicates that there is a significant demand for the services that the sponsor intends to offer, the feasibility consultant is often involved in financial planning by assisting management in the development of a baseline financial model. This model is used by management to evaluate the potential financial viability of the proposed community and to generate a variety of sensitivity analyses. Sensitivity analyses allow management to assess the risks associated with several of the more significant assumptions, including monthly service fee and entrance fee assumptions, construction costs, interest rates, or operating expenses.

Preliminary market analyses and financial planning services may be accomplished by the feasibility consultant as a service to management and may or may not require the issuance of a report. Should the feasibility consultant be asked to present a report for internal use only, it may take the form of an assembly report or a compilation report. In many cases this information is in a partial presentation format and can be characterized as projected information, since many of the assumptions are hypothetical at this stage of planning.

The next level of services that the feasibility consultant offers concerns the regulatory approval process. Each state has various regulatory processes for the approval of nursing beds, assisted living units, continuing care contracts, and so on. Most of these regulatory processes require prospective financial statements with a consultant's report as part of the regulatory filing. Also, the document submission requirements generally require the gathering of a variety of market, financial, capital, and operational data, which must be tabulated and summarized in a variety of ways. Feasibility study consultants are often engaged to assist in this process because they are generally familiar with the state requirements and may be able to assist in expediting the process. The feasibility study consultant may issue a compilation, examination, or full feasibility study report in connection with the regulatory process. Often these reports will be limited in distribution to the regulatory agency.

The most extensive service that a feasibility study consultant offers is the preparation of a full feasibility study report for financing purposes. This service generally begins after the planning efforts have been completed by management, premarketing of the facility has occurred at some level, design is complete, construction costs are firm, and the financing plan is in place at specific estimated rates. Full service includes an examination of key management assumptions and the presentation of the financial forecast with the formal development of the feasibility study report. A full feasibility study report is a requirement for most public bond offerings, and the feasibility consultant participates as a member of the financing team when a public bond offering is conducted.

Feasibility Study Process

A full financial feasibility study, intended for financing purposes, is conducted in seven steps:

1. Selection of the feasibility consultant.
2. Information gathering and preliminary discussions.
3. Market analysis.
4. Examination procedures.
5. Financial analysis.
6. Report preparation.
7. Review of offering statement and other legal documents.

The tasks involved in each step are discussed below.

Step 1: Selection of the Feasibility Consultant

The feasibility study process begins with management's selection of a feasibility consultant. Wise selection criteria may include the consultant's

- Familiarity with the local market.
- Familiarity with the products and services.
- Experience in the industry.
- Experience with the feasibility process.
- Depth and breadth of staff.
- Timing.
- Pricing.

One or more proposals are typically solicited and reviewed by management and a committee from the board of directors. Many long-term care communities conduct a two-step proposal process; the first step requests a written proposal, and the second step requests an on-site interview. Key items to consider throughout the proposal period include which individuals actually will do the work and their experience in providing services. Also, management should understand all proposed charges, including any expected additional or out-of-pocket expenses. Whereas all financing team services are price-sensitive, the risk of involving the wrong professionals can be substantial. Accordingly, management's experience with the consultant is often the overriding selection criterion.

Step 2: Information Gathering and Preliminary Discussions

Once the feasibility consultant is chosen, the next step involves collection of the data and information required by the feasibility consultant to complete the task. The request for information typically includes a wide variety of financial, market, construction, and financing information. Table 8–3 presents a feasibility consultant's typical data request list for a long-term care development.

Although management is ultimately responsible for the assumptions and the financial statements within the report, the initial data gathering and formulation of assumptions usually involve a wide variety of professionals. An investment banker is responsible for providing the financing assumptions; an architect or contractor, for the construction costs; an actuary, for the population flow; and a marketing firm, for the pricing, fill-up, and occupancy information. Gathering the information in an efficient and effective manner at the beginning of the process makes for a smoother feasibility study. It is often valuable to have a kickoff meeting, with all of the professionals present, to review this initial set of data and to thoroughly discuss the project.

Step 3: Market Analysis

The market analysis requires management to determine the facility's primary market area, the geographic area from which management anticipates a significant number of its residents will originate. Obtaining the ZIP codes or cities of residence of the depositors or current residents before they moved into the long-term care community is an efficient way to analyze the community's primary market area. If this information is not available early in the process, local market factors may be considered and subsequently confirmed by presale data.

Once the primary market area has been determined, the feasibility consultant obtains a variety of demographic, economic, and real estate

TABLE 8-3 Typical Data Request List by Feasibility Consultant

Background information
- History of facility—organizationally and product and service structure
- Number of corporate entities and management and board structure
- Historical financial statements and other selected operating data and statistics

Project data
- Project team list and background
- Description of project, including product and service definition
- Location and size of site
- Estimated timeline for financing, construction and fill-up
- Description of marketing efforts to date and presale list of units
- Preliminary sources and uses of funds
- Estimates of development costs
- Estimates of costs for zoning, licensing, utilities, and other regulatory compliances

Construction data
- Estimates of construction costs
- Timing of construction
- Payout schedule
- Estimates of other construction-related costs
- Estimates of furniture and fixture costs
- Estimates of architectural and engineering costs
- Estimates of contingency costs
- Estimates of the useful life of the construction and related costs
- Estimates of construction period interest expense and interest earnings

Financing costs
- Description of financing plan
- Estimates of total financing package
- Estimates of interest rates
- Loan amortization schedule
- Estimates of soft costs
- Estimates of trustee funds and funding requirements

(continued)

TABLE 8-3 Typical Data Request List by Feasibility Consultant (concluded)

Operating revenues (for existing and new products and services)
- Actuarial estimates of turnover of housing units and mortality and morbidity rates
- Entrance contracts and structure
- Rate structure
- Fill-up estimates
- Occupancy estimates
- Revenue estimates
- Entrance fee amortization
- Ancillary services estimates
- Interest earnings estimates
- Contribution estimates

Operating expenses (for existing and new products and services)
- Staffing plan, including estimates of FTEs by department
- Salary and wage scale
- Employee benefits plans and rates
- Raw-food costs
- Utility costs
- Medical supplies
- Contracted services
- Other operating expenses

---

information for the primary market area. The economic and real estate information is used to more fully understand the market in which the retirement community will operate. Demographic data determine the number of individuals who would be qualified, in terms of age, income, and assets, to move into the retirement community.

For age-qualified individuals, again the profile of current residents or depositors is used. Usually, the feasibility consultant will analyze the population of individuals aged 65 and older in the primary market.

To determine the income-qualified individuals, the feasibility consultant will initially use the specifications of the retirement community. Many communities assume that an individual with a monthly income of 1.75 or 2 times the applicable monthly service fees would be income-qualified for a particular housing unit. The feasibility consultant will test this assumption by analyzing the services covered under the community's resident agreement and adding to the monthly service fees other expenses that the resident might reasonably incur. Similar estimates are made for assisted living and nursing services.

Many retirement communities require the payment of an entrance fee on admission. The feasibility consultant therefore needs to ascertain the number of individuals who are likely to have enough funds to afford the entrance fee. A surrogate, such as the percentage of the targeted age-qualified population who own a home, may be used for determining asset qualification. Such surrogate data may be based on the reasonable assumption that the equity from the sale of an individual's home could be applied toward the payment of the entrance fee.

After completing the analysis of the age-, income-, and asset-qualified individuals, the feasibility consultant focuses on analyzing the competitive environment in which the community operates. This process involves identifying all similar retirement communities and collecting pertinent information about those communities, including:

- Name
- Ownership
- Number of units
- Configuration of units
- Occupancy
- Services provided
- Expansion plans
- Entrance fee structure
- Monthly fee structure

This information is usually collected via both phone interviews and site visits. Once the competitive matrix has been completed, the feasibility consultant is ready to calculate the market penetration rate for the retirement community.

There are generally two separate market penetration rates calculated for each product (housing, assisted living, and nursing care):

- *Gross penetration rate*, expressed as a percentage of the ratio of total units in the market, including all planned units, compared with the age-, income-, and asset-qualified relevant population.

- *Net penetration rate,* expressed as a percentage of the number of individuals required to fill all of the facility's new units and competitors' new units in the primary market area, compared with the number of age-, income-, and asset-qualified relevant population in the primary market.

The gross penetration rate depicts the overall level of saturation of the primary market for the specific product measured. The net penetration rate indicates the capture rate that must be achieved from the primary market, excluding those individuals already living in the specified product, to fill the new product in the market. The ability to penetrate a certain percentage of the market varies based on the geographic location of each facility and the type of product. This variance is primarily due to the different levels of acceptance that retirement housing has achieved in different areas of the country. Table 8–4 illustrates a sample calculation of gross and net market penetration rates.

The gross and net penetration rates should serve as an indicator in estimating the potential market available for the independent living units at the retirement community, but penetration must be considered in conjunction with other market factors, such as occupancy levels at existing competitive retirement communities within or near the primary market area, the number of proposed facilities within the primary market area, and the marketing plans and efforts of management.

Additional data typically gathered during the market study include real estate sales activity, including data on the average selling price of homes in the market and the average number of days that a home remains on the market. This information is important in the estimation of the affordability of the entrance fee, since many individuals will sell their home and use the proceeds to fund the entrance fee. Also, the strength of the market will indicate prospective residents' ease in selling their homes and moving to the facility. A slow market can delay move-ins to the facility.

Feasibility report consultants also typically compile and review a list of major employers in the area and their status and stability in the market. The economic activity of major employers and the employment situation in general influence the housing market and the stability of the local market area.

All projects that involve market-rate housing require the sponsor to premarket a substantial portion of the units, currently in the 70

TABLE 8-4 Sample Market Penetration Calculations

Gross market penetration calculation	
Gross number of existing independent living units (ILUs)	420
New units, including project	150
Total new and existing units	570
Qualified individuals aged 75 and over	2,500
Gross penetration rate	22%
Net Market Penetration Calculation	
New units, including project	150
Qualified individuals aged 75 and over	2,500
Less current filled ILUs	− 420
Net qualified individuals	2,080
Net penetration rate	7%

percent range. This process represents a form of market research, since individuals are generally quoted prices and asked to select entrance fee programs, services, and specific living units. Much can be learned by the sign-up patterns during the presale effort and by the speed at which the presales are accomplished.

The qualification for a presale of a unit includes the collection of a deposit (at least a 10 percent of the total entrance fee) and a signed entrance agreement contract. In most cases there is also a cancellation provision, with a full or partial refund of the deposit permitted in accordance with most state regulations.

The feasibility consultant confirms the presale list through a formal survey process and solicits various data elements from the survey as well as the application submitted for admission to the facility. Information gathered generally includes:

- Asset and income information.
- Current address and living arrangement.
- Reasons for selecting the facility.
- Estimated timing of move-in.
- Need to sell current home.
- Deposits made at other retirement communities.

Assessment of the overall market demand for housing, assisted living, and nursing care is more of an art than a science. The information enumerated above—the presale list, the number of qualified individuals in the market, market penetration rates, competitor status and plans, real estate market, and general economic conditions—is all relevant, but the final conclusions of market demand must be evaluated in light of the consultant's experience and overall comfort with the marketplace.

Management uses the market study to confirm the strength of the market and to estimate or confirm its fill-up rates and anticipated occupancy rates, unit configuration, and pricing. If the feasibility consultants gather and assemble the data themselves, they will rely on the market study as the examination process for the market evaluation. Otherwise, the feasibility consultants will perform separate examination procedures to verify the information from management.

Step 4: Financial Analysis

A financial analysis is prepared either by management or by the consultant using a microcomputer spreadsheet application.

Typically the consultant will have a variety of structured models that incorporate the various data requirements associated with long-term care and retirement housing products and services. Such models facilitate the rapid entry of data and customization for the project at hand, allowing a much more timely and accurate process than management could accomplish by setting up a model from scratch. It is also normally more efficient and supportive of the financing process for the consultant to perform the modeling, updates, and sensitivity analysis, unless management has an experienced staff with these tools in place.

The financial model requires information on financial and operational assumptions and data gathered from management and other professionals, including:

- Construction costs, timing, and cash flows.
- Financing costs, amortization schedule, financing rate, principal, interest earnings, and trustee funds.

- Sources and uses of funds for the project.
- Budget information for current operations, if any.
- Balance sheet information for existing operations.
- Working capital assumptions.
- Actuarial assumptions on turnover, morbidity, and mortality.
- Occupancy, fill-up, and rate assumptions.
- Entrance fee collection and amortization assumptions.
- Operating cost assumptions, including analysis of full-time equivalents (FTEs).
- Interest cost, depreciation, and amortization expense.

All of these financial and operating data are entered into the computer model, and a baseline financial forecast is prepared for the project and any existing operations. Because of the overall credit evaluation performed by the financing team and the financiers, all operations of an entity are included within the baseline model for a financing forecast. Depending on the needs of management and of the financing team, the project may or may not be modeled separately, in order to illustrate the financial feasibility of the project itself.

The baseline model is reviewed for accuracy of input and accuracy of assumptions by the consultant, management, and selected members of the financing team, including the underwriter. In most cases the assumptions are fine-tuned and several iterations of the model are processed and distributed before the baseline is finalized. In fact, the baseline model will continue to evolve throughout the financing period as the costs and timing of construction, the costs and timing of financing, and other key assumptions are adjusted and finalized. Each change in baseline assumptions requires a rerun of the model.

Sensitivity analyses are prepared subsequent to the baseline model. These analyses are intended to provide information to the finance team and, ultimately, to the financiers regarding how the financial position, results of operations, and cash flows will be affected by variations in key assumptions. An example of a sensitivity analysis would be to show the financial impact of a 1 percent decline in overall occupancy or of an extension of the fill-up period beyond the baseline forecast.

Step 5: *Examination Procedures*

Once management's assumptions have been fully developed, the feasibility consultant applies examination procedures to confirm the reasonableness of key assumptions and the presentation of the financial statements. Various types of examination procedures are applied to the forecast to support the basis for key assumptions, including:

- *Confirm data and information with outside parties* to obtain an independent third party's support for the assumptions, such as confirmation of validity of presales with outside prospective residents; confirmation of the financing plan with the underwriter; receipt of signed construction contracts from construction company; receipt of signed architectural contracts; and copies of deeds or land purchase options, regulatory approvals, development agreements and management agreements, and actuarial studies to support actuarial estimates.
- *Conduct testing procedures of client records* to vouch certain balances and data to management's assumptions and the computer model, such as a review of the sign-up list, and trace to applications and marketing materials for rates and entrance fees; trace project-to-date expenditures to accounting records and to original invoices from vendors; trace balance sheet accounts to accounting records and underlying detail support such as accounts receivable ledgers and accrued expense ledgers; trace balance sheet accounts to audited financial statements; vouch FTEs and salary and wage rates to payroll records for existing facilities or to local market wage indexes for start-up facilities; trace employee benefits estimates to historical details and current-year budget estimates; trace raw-food, utilities, medical supplies, and other operating costs to historical levels for existing facilities; trace revenue estimates to price lists approved by the board of directors and occupancy estimates to historical performance data; trace entrance fee estimates and unit rates to marketing materials; recompute depreciation estimates and trace useful life estimates to historical estimates or industry estimates; and trace interest cost, debt service payments, and trustee account balances and funding requirements to the loan documents.

- *Perform analytical procedures* to test the overall reasonableness of operating revenues and operating expenses, including the following: compare FTEs by department with industry averages for the specific products and services provided; compare the wage scale to local-market averages; compare the benefits package with local-market averages and industry averages; compare cost-per-meal with industry averages; compare utility cost-per-square foot with local-market rates and industry averages; and perform other selected analytical comparisons as deemed appropriate, such as revenue-per-day and cost-per-day by product line, current-year budget to historical figures and current-year actual figures, and revenue and expense forecasts to industry averages.

Step 6: Report Preparation

After completion of the market study, financial analysis, and examination procedures, the consultant prepares a formal feasibility study report. The report consists of these three sections:

1. Feasibility study opinion letter.
2. Financial statements.
3. Summary of significant accounting policies and assumptions.

The feasibility study opinion letter is a form letter issued in a format prescribed in the AICPA's *Guide for Prospective Financial Information*. Table 8–5 presents the standard opinion letter format for a full feasibility study report as well as for examination, compilation, agreed-upon procedures, and assembly reports. The forms may be modified to some degree in describing the project and the financing, but the opinion section is a standard requirement without modification. The consultant signs and dates the feasibility study report letter. 8–5

The financial statements contained within the feasibility study are assembled by management or the consultant from the computer model. The format of the financial statements should be consistent with that of the historical financial statements for the facility or with industry standards for a new entity. This format will facilitate both historical comparisons and future comparisons. The forecasted financial statements should be limited to a maximum five-year forecast period. A special financial statement, not normally included in historical financial statements, is typically included with the feasibility study. This

TABLE 8–5 Standard Opinion Letter Formats for Feasibility Studies

Feasibility study report

a. The Board of Directors
Example Hospital
Example, Texas

b. We have prepared a financial feasibility study of Example Hospital's plans to expand and renovate its facilities. The study was undertaken to evaluate the ability of Example Hospital (the hospital) to meet the hospital's operating expenses, working capital needs, and other financial requirements, including the debt service requirements associated with the proposed $25,000,000 (legal title of bonds) issue, at an assumed average annual interest rate of 10 percent during the five years ending December 31, 19__.

c. The proposed capital improvements program (the program) consists of a new two-level addition, which is to provide 50 additional medical-surgical beds, increasing the complement to 275 beds. In addition, various administrative and support service areas in the present facility are to be remodeled. The hospital administration anticipates that construction will begin June 30, 19__, and will be completed by December 31, 19__.

d. The estimated total cost of the program is approximately $30,000,000. It is assumed that the $25,000,000 of revenue bonds that the Example Hospital Finance Authority proposes to issue would be the primary source of funds for the program. The responsibility for payment of debt service on the bonds is solely that of the hospital. Other funds necessary to finance the program are assumed to be provided from the hospital's funds, from a local fund drive, and from interest earned on funds held by the bond trustee during the construction period.

e. Our procedures included analysis of:
- Program history, objectives, timing, and financing
- The future demand for the hospital's services, including consideration of
  Economic and demographic characteristics of the hospital's defined service area
  Locations, capacities, and competitive information pertaining to other existing and planned area hospitals
  Physician support for the hospital and its programs
  Historical utilization levels
- Planning agency applications and approvals
- Construction and equipment costs, debt service requirements, and estimated financing costs
- Staffing patterns and other operating considerations
- Third-party reimbursement policy and history
- Revenue/expense/volume relationships.

f. We also participated in gathering other information, assisted management in identifying and formulating its assumptions, and assembled the accompanying financial forecast based on those assumptions.

(continued)

TABLE 8-5 Standard Opinion Letter Formats for Feasibility Studies *(continued)*

g. The accompanying financial forecast for the annual periods ending December 31, 19__ through 19__, is based on assumptions that were provided by, or reviewed with and approved by, management. The financial forecast includes:
   - Balance sheets
   - Statements of revenues and expenses
   - Statements of cash flows
   - Statements of changes in fund balance

h. We have examined the financial forecast. Our examination was made in accordance with standards for an examination of a financial forecast established by the American Institute of Certified Public Accountants and, accordingly, included such procedures as we considered necessary to evaluate both the assumptions used by management and the preparation and presentation of the forecast.

i. Legislation and regulations at all levels of government have affected and may continue to affect revenues and expenses of hospitals. The financial forecast is based on legislation and regulations currently in effect. If future legislation or regulations related to hospital operations are enacted, such legislation or regulations could have a material effect on future operations.

j. The interest rate, principal payments, program costs, and other financing assumptions are described in the section entitled "Summary of Significant Forecast Assumptions and Rationale." If actual interest rates, principal payments, and funding requirements are different from those assumed, the amount of the bond issue and debt service requirements would need to be adjusted accordingly from those indicated in the forecast. If such interest rates, principal payments, and funding requirements are lower than those assumed, such adjustments would not adversely affect the forecast.

k. Our conclusions are presented below:
   - In our opinion, the accompanying financial forecast is presented in conformity with guidelines for presentation of a financial forecast established by the American Institute of Certified Public Accountants.
   - In our opinion, the underlying assumptions provide a reasonable basis for management's forecast. However, there will usually be differences between the forecasted and actual results, because events and circumstances frequently do not occur as expected, and those differences may be material.
   - The accompanying financial forecast indicates that sufficient funds could be generated to meet the hospital's operating expenses, working capital needs, and other financial requirements, including the debt service requirements associated with the proposed $25,000,000 bond issue, during the forecast periods. However, the achievement of any financial forecast is dependent on future events, the occurrence of which cannot be assured.

l. We have no responsibility to update this report for events and circumstances occurring after the date of this report.

(continued)

TABLE 8-5 Standard Opinion Letter Formats for Feasibility Studies *(continued)*

Examination report

We have examined the accompanying forecasted balance sheet and the related forecasted statements of income, retained earnings, and cash flows of XYZ Company as of December 31, 19__, and for the year then ending. Our examination was made in accordance with standards for an examination of a forecast established by the American Institute of Certified Public Accountants and, accordingly, included such procedures as we considered necessary to evaluate both the assumptions used by management and the preparation and presentation of the forecast.

In our opinion, the accompanying forecast is presented in conformity with guidelines for presentation of a forecast established by the American Institute of Certified Public Accountants, and the underlying assumptions provide a reasonable basis for management's forecast. However, there will usually be differences between the forecasted and actual results, because events and circumstances frequently do not occur as expected, and those differences may be material. We have no responsibility to update this report for events and circumstances occurring after the date of this report.

Compilation report

We have compiled the accompanying forecasted balance sheet and the related forecasted statements of income, retained earnings, and cash flows of XYZ Company as of December 31, 19__, and for the year then ending, in accordance with standards established by the American Institute of Certified Public Accountants.

A compilation is limited to presenting in the form of a forecast information that is the representation of management and does not include evaluation of the support for the assumptions underlying the forecast. We have not examined the forecast and, accordingly, do not express an opinion or any other form of assurance on the accompanying statements or assumptions. Furthermore, there will usually be differences between the forecasted and actual results, because events and circumstances frequently do not occur as expected, and those differences may be material. We have no responsibility to update this report for events and circumstances occurring after the date of this report.

Agreed-upon procedures report

Board of directors—XYZ Corporation

Board of directors—ABC Company

At your request, we have performed certain agreed-upon procedures, as enumerated below, with respect to the forecasted balance sheet and the related forecasted statements of income, retained earnings, and cash flows of DEF Company, a subsidiary of ABC Company, as of December 31, 19__, and for the year then ending. These procedures, which were specified by the boards of directors of XYZ Corporation and ABC Company, were performed solely to assist you in connection with the proposed sale of DEF Company to XYZ Corporation. It is understood that this report is solely for your information and should not be used by those who did not participate in determining the procedures.

(continued)

TABLE 8-5 Standard Opinion Letter Formats for Feasibility Studies *(concluded)*

a. With respect to forecasted rental income, we compared the assumptions about expected demand for rental of the housing units to demand for similar housing units at similar rental prices in the city area in which DEF Company's housing units are located.

b. We tested the forecast for mathematical accuracy.

Because the procedures described above do not constitute an examination of prospective financial statements in accordance with standards established by the American Institute of Certified Public Accountants, we do not express an opinion on whether the prospective financial statements are presented in conformity with AICPA presentation guidelines or on whether the underlying assumptions provide a reasonable basis for the presentation.

In connection with the procedures referred to above, no matters came to our attention that caused us to believe that rental income should be adjusted or that the forecast is mathematically inaccurate. Had we performed additional procedures or had we made an examination of the forecast in accordance with standards established by the American Institute of Certified Public Accountants, matter might have come to our attention that would have been reported to you. Furthermore, there will usually be differences between the forecasted and actual results, because events and circumstances frequently do not occur as expected, and those differences may be material. We have no responsibility to update this report for events and circumstances occurring after the date of this report.

Assembly report

We have assembled, from information provided by management, the accompanying forecasted balance sheet and the related forecasted statements of income, retained earnings, and cash flows of XYZ Company as of December 31, 19__, and for the year then ending. (This financial forecast omits the summary of significant accounting policies.) We have not compiled or examined the financial forecast and express no assurance of any kind on it. Further, there will usually be differences between the forecasted and actual results, because events and circumstances frequently do not occur as expected, and those differences may be material. In accordance with the terms of our engagement, this report and the accompanying forecast are restricted to internal use and may not be shown to any third party for any purpose.

Source: American Institute of Certified Public Accountants. *Guide for Prospective Financial Information.* Washington, D.C.: American Institute of Certified Public Accountants, 1991, pp. 25–30, 41–43, 51.

statement computes one or more financial ratios that are important for the review of the financing team and the financiers. The ratios generally include a debt service coverage ratio, computed as the ratio of funds available for debt service to the annual or maximum annual debt service on the long-term debt. This ratio is an indicator of the strength of operations to produce cash funds for payment of debt service.

Other financial ratios often presented with the debt service coverage ratios include the ratio of cash and investments of the organization compared with maximum annual debt service and the ratio of cash and investments of the organization compared with the long-term debt outstanding. Both of these ratios are intended to illustrate the liquidity position of the organization and the availability of cash funds for payment of debt service. A rule of thumb often applied is a ratio of cash and investments to the long-term debt outstanding in the range of 0.3, which indicates that cash and investments on hand would amount to approximately 30 percent of the long-term debt balance. The debt service coverage ratio and the liquidity ratios are often presented in the format shown in Table 8–6.

Table 8–6   Sample Facility Debt Service Coverage and Liquidity Ratios for the Year Ending December 31, 199X

Debt service coverage ratio	
Net income	$ 500,000
Add depreciation and interest expense	1,500.000
Add entrance fees collected	600,000
Less amortization of entrance fees	(500,000)
Funds available for debt service	$ 2,100,000
Maximum annual debt service on long-term debt	$ 1,300,000
Debt service coverage ratio	1.61
Ratio of cash and investments to maximum annual debt service	
Cash and investments	$ 4,000,000
Maximum annual debt service on long-term debt	$ 1,300,000
Ratio of cash and investments to maximum annual debt service	3.07
Ratio of cash and investments to long-term debt	
Cash and investments	$ 4,000,000
Long-term debt	$12,000,000
Ratio of cash and investments to long-term debt	0.33

Following the forecasted financial statements within the feasibility study report is a section describing the accounting policies used by the organization and a summary of the key assumptions underlying the preparation of the forecast. The sequence and types of disclosures are generally as follows:

- Introduction and statement of responsibility.
- Description of corporation and project.
- Sources and uses of funds (see example, Table 8–7).
- Market analysis.
- Utilization assumptions.
- Accounting policies.
- Balance sheet assumptions.
- Inflation assumptions.
- Revenue assumptions.

Table 8–7 Sample Disclosure of Sources and Uses of Funds

Sources of funds	
Long-term debt	$12,000,000
Interest earnings	500,000
Equity contribution	2,500,000
Total sources of funds	$15,000,000
Uses of funds	
Land	$ 500,000
Construction costs	8,500,000
Furniture and fixtures	400,000
Architect and engineering	600,000
Total construction and related costs	$10,000,000
Development costs	$500,000
Marketing costs	700,000
Working capital	400,000
Funded interest	1,500,000
Debt service reserve fund	1,200,000
Financing costs	700,000
Total uses of funds	$15,000,000

- Expense assumptions.
- Cash flow assumptions.
- Sensitivity analysis.

The notes to the report may be written by management or by the consultant; however, it should be made explicit that the assumptions are the responsibility of management and not the consultant. This is a very important issue, since the buy-in and capacity of management to execute its plans in accordance with the assumptions provided is the cornerstone of the forecast. The consultant should not issue a feasibility study report in circumstances in which management does not have the requisite skills and experience to execute the development, marketing, construction, and operations of the project.

Sensitivity analyses should be presented in the notes to the report for selected assumptions that are material to the forecast and are particularly sensitive to future variations. A sensitivity analysis assists the user of the report to understand the potential impact that variations in assumptions may have on the achievability of the forecast. Sensitivity analyses are often presented for the following assumptions:

- Fill-up period
- Occupancy level
- Rate structure
- Payer mix
- Inflation levels
- Staffing levels

An example of a sensitivity analysis for occupancy level is shown in Table 8–8. This sensitivity analysis reflects the impact on funds available for debt service and the debt service coverage ratio for a 1 percent decline in the overall occupancy rate of the facility. It also presents the occupancy level at which the debt service coverage level hits breakeven 1.00.

Step 7: Review of Offering Statement and Other Legal Documents

Public bond offerings are sold through an offering statement prepared by the underwriter for the bond issue. The offering statement generally describes the following types of information:

Table 8-8 Sample Facility Occupancy Sensitivity Analysis for the Year Ending December 31, 199X

	Forecasted Data	1% Decline Analysis	Breakeven Analysis
Occupancy level	95%	94%	85%
Net income	$ 500,000	$ 415,000	$ (300,000)
Funds available for debt service	$2,100,000	$2,015,000	$1,300,000
Maximum annual debt service	$1,300,000	$1,300,000	$1,300,000
Debt service coverage ratio	1.61	1.55	1.0

- Financing authority.
- Parties to the transaction.
- Terms of the financing.
- Security.
- Facility.
- Project.
- Market.
- Financial performance.
- Risks.
- Feasibility study.
- Historical financial statements.

The feasibility study is generally incorporated into the offering statement as an appendix. Also, the underwriter usually will extract certain information (including sources and uses of funds, debt service coverage schedule, project description, and market information) from the feasibility study and incorporate the information within the text of the offering statement.

The feasibility study consultant has a responsibility to review the offering statement and to ascertain that the information presented by the underwriter is consistent with the feasibility study. Also, the consultant will want to determine that the responsibility of management for the forecast and assumptions is clearly defined and that undue responsibility is not assigned to the consultant.

The underwriter will require a consent from the feasibility consultant to include the feasibility study report within the offering statement. The consultant should conduct a review of the final offering statement before issuing such consent. Also, the consent is generally dated consistent with the offering statement, which requires the consultant to perform a certain level of due diligence between the date of the feasibility study report and the consent date.

Due diligence procedures often include the following:

- Read the latest available interim financial statements, operation reports, and any relevant prospective information, such as budgets; consider the prospective results in relation to the actual results achieved in the interim period; and inquire whether the accounting principles used in the preparation of such information are consistent with the principles used in preparing the forecast.
- Read the entire prospectus and other pertinent portions of the registration statement and consider that information in relation to prospective results and the summary of significant assumptions.
- Inquire of and obtain written representations from the responsible party, including those individuals responsible for matters significant to the financial forecasts, as to whether there are any events, plans, or expectations that may require the financial forecast to be modified or that should be disclosed, so that the forecast will reflect the responsible party's judgment based on present circumstances of the expected conditions and its expected course of action. In lieu of obtaining an additional representation letter at the consent date, the accountant may have the responsible party update the representation letter originally signed at the report date.
- Read the available minutes of meetings of the board of directors and related committees. Regarding meetings for which minutes are not available, inquire about matters dealt with at such meetings.
- Make such additional inquiries or perform such procedures as considered necessary and appropriate to dispose of questions that arise in carrying out the foregoing procedures.

Several other legal documents are incorporated into a tax-exempt bond offering, including a trust indenture and a loan agreement. The consultant should review these documents to ascertain that the terms of the financing are consistent with the forecast and disclosures.

Summary

The financial feasibility study represents the most crucial and perhaps the most trying stage in the life of a new long-term care development. Of paramount importance are the skillful coordination of professionals from many disciplines, the careful formulation and testing of key planning assumptions, the performance of all study tasks with analytical rigor, and reporting in full compliance with accepted standards and principles. When a feasibility study is properly designed, managed, and executed, with expert guidance by an experienced feasibility specialist, the process will yield a wealth of useful information for sponsors and developers and will make a persuasive argument for prospective investors and other interested parties.

Resources

In researching this project and conducting my usual client work, I have discovered long-term care business publications, organizations, and events too numerous to list. A few especially rich stores of information, assembled and annotated here, will provide all the basic principles, expert advice, and fresh news the long-term care investor needs to get started.

Investment Opportunities and News

The investment community keeps a close eye on long-term care these days. Hundreds of lenders, venture capitalists, advisors, and facility owners and operators attend the National Investment Conference (NIC) for the Senior Living and Long Term Care Industries, a nonprofit educational forum held each October in Washington, D.C., to discuss new directions and investment opportunities in the field. Conference proceeds are used to fund research and other efforts to stimulate capital formation in the senior living and long-term care industries. Besides serving as an essential information clearinghouse and speaker's forum, the NIC hosts the long-term care industry's real movers and shakers and every year provides the backdrop for key business deals. Contact the NIC organizers, headquartered in Annapolis, Maryland, at 410/267-0504.

Amid the glut of health care periodicals, McKnight's *Long-Term Care NEWS* is required reading for anyone following the long-term care industry. Every monthly issue covers new federal and state legislation and regulations, long-term care stock performance, company earnings and transactions, construction reports, product and supplier notes, nursing and pharmacy perspectives, and pointed observations and commentaries.

Strategic and Financial Planning

The social phenomenon we dub "population aging," remember, comprises living, thinking, feeling beings and their dedicated caregivers. Nancy Alfred Persily's *Eldercare: Positioning Your Hospital for the Future* (Chicago: American Hospital Publishing, 1991) is a fine introduction to the business and, more important, the human challenges of providing comprehensive healthcare services for the elderly. Persily explores the intricacies of long-term care delivery; examines key trends and successful practices; and urges planners, administrators, and program directors to make changes and seize opportunities. Illustrative case studies round out this excellent primer.

A rapidly growing population of patients who require posthospital (subacute) long-term care in freestanding facilities, at home, and in outpatient settings seriously challenges our notion of an integrated care continuum. *Outpatient Case Management: Strategies for a New Reality* (Chicago: American Hospital Publishing, 1994), edited by Michelle Regan Donovan and Theodore A. Matson, is an exhaustive treatment of the subacute care revolution and the critical linking mechanisms that must be developed to contain our sprawling continuum of care. This seminal book includes a thoughtful introduction to case management principles and practices as well as an exhaustive treatment of working models, including long-term care, rehabilitation, and other extended care case management programs for outpatients. An annotated bibliography includes several listings covering case management in aging, long-term care, and home care service systems.

Hospitals and other healthcare organizations planning to develop facilities or expand their long-term care offerings will first want to test their financial wherewithal and creditworthiness. In *The Financially Competitive Healthcare Organization* (Chicago: Probus Publishing, 1994), Kenneth Kaufman and Mark Hall dispense timely and timeless advice about sound financial and capital planning. Building from a thesis espousing truly informed, decisive executive leadership and a close familiarity with the needs and preferences of the capital markets, the book excels in describing basic principles as well as the intelligent application of sophisticated new financial modeling techniques. Long-term care investors will find the chapters on acquisition analysis and project sizing especially enlightening.

Of course, the local cost of healthcare labor and the effective management of human resources will figure prominently in any long-term care investment, but the fair treatment of such complex issues would require another book altogether. Planners will have their homework and legwork cut out for them. Labor costs do vary from one market to the next, and mounting efforts to unionize healthcare professionals from all walks will only complicate matters. Your state nursing home association likely can provide area-specific salary ranges for nurse directors and other top positions, but don't expect to get the whole labor picture from any one source. Regarding how best to recruit, retain, train, and manage healthcare professionals of all disciplines and specialties, long-term care facility operators and managers surely will rely on their own experiences, favorite publications, trusted advisors, and time-tested wisdom to guide them.

Author Biographies

About the Editor

Ronald E. Mills is an independent healthcare writer and editor. For 12 years Mr. Mills has served the industry's top executives, consultants, and publishers, first on the marketing team at Chi Systems, then as healthcare editor for Ernst & Young's Chicago office and Midwest region, and since 1993 as a free-lancer in national practice based in Highland Park, Illinois. His clients, ranging from Fortune 500 companies to solo practitioners, include managed care organizations, business strategists, financial advisors, facility planners, pharmacoeconomists, systems and equipment vendors, professional associations, public relations and advertising agencies, and book and journal publishers. Mr. Mills earned his bachelor's and master's degrees in English Language and Literature from the University of Michigan, Ann Arbor. He lives with his wife Laura and their son Daniel in Evanston, Illinois.

About the Authors

Roy W. Byers, CPA, is a partner of KPMG Peat Marwick LLP based in Harrisburg, Pennsylvania. With over 20 years' consulting experience in healthcare and real estate, Mr. Byers has prepared financial feasibility studies and supervised the overall planning, development, and financing of numerous successful projects, including proprietary, nonprofit, and government-owned and -operated facilities housing congregate living, continuing care, skilled nursing, and other long-term care programs. He received his BA in Economics and MA in Business Administration from the University of Pittsburgh.

Kathleen R. Crampton is vice president of marketing for HealthTech Services Corporation, Northbrook, Illinois. A longtime leader in community health, Ms. Crampton was previously a consultant with Coopers & Lybrand and Kaiser Permanente, chief financial officer of the District of Columbia Institute for Mental Health, director of planning and development for the

Harvard Community Health Plan, and director of Ambulatory Services at Children's Hospital Medical Center, Boston. She earned a bachelor's degree in History and a master's degree in Urban Studies from Boston University, a master of public health (MPH) degree in Health Administration from the Harvard School of Public Health, and an MBA in Finance from George Washington University.

Christopher J. Donovan, JD, is a partner of McDermott, Will & Emery based in Boston and practices in all areas of the subacute care, long-term care, and assisted living industries. Mr. Donovan represents providers, lenders, and sponsors in connection with the financing, permitting, and development of subacute care facilities, nursing homes, assisted living facilities, ambulatory surgery centers, clinics, age-restricted housing, and related facilities. He is a member of the Boston and Massachusetts Bar associations; received his BSFS, magna cum laude, from Georgetown University; and studied law at Boston College, where he was the note and comment editor for the *Boston College International and Comparative Law Review*.

Marietta Donne Hance is a senior consultant with the health care information technology practice of Ernst & Young LLP based in Chicago. Ms. Hance's health care experience features information systems selection and implementation, operations analyses, and work process redesign in a variety of clinical settings. She earned her bachelor's degree in nursing from Purdue University and an MBA from Loyola University.

Darolyn Jorgensen is a candidate for the master of health services administration degree at George Washington University, with a special interest in long-term care. She has served a number of internships and assistantships, including a 1994 internship for Senator Edward Kennedy and the U.S. Senate Committee on Labor and Human Resources, Health Policy Office. Ms. Jorgensen earned a bachelor's degree in Urban Studies and Planning, with minors in Economics and Latin American Studies, from the University of California, San Diego. She also studied at the Institute for Social and International Studies in Barcelona, Spain.

Stephen B. Kaufman is founder and vice chairman of HealthTech Services Corporation. With more than 20 medical patents and 50 manufacturing processes to his credit, Mr. Kaufman has spent his entire career devising innovative solutions to healthcare and business challenges. HealthTech sprang from his invention of the Supported Independence™ system package, which features automated home nursing assistants linked to a central station providing round-the-clock patient monitoring. Mr. Kaufman was previously director of advanced technology at Baxter Healthcare Corporation and devised the concept for Baxter's automated pharmacy system. He has consulted to business and government entities around the world on technology matters.

P. Andrew Logan is a manager with Zelenkofske Axelrod's Health Care Services Group based in Atlanta. With concentrated experience in assisted living and long-term care, Mr. Logan brings his financial and management expertise to healthcare facility expansion, refinancing, and bond offerings. He specializes in conducting market and financial feasibility studies as well as management and operational reviews to help assisted living and long-term care facilities maximize revenues. Mr. Logan received a bachelor's degree in Psychology and an MBA in General Management from the University of Virginia.

Michael A. Lucas is a principal of O'Toole, Lucas & Associates, a real estate appraisal and consulting firm based in Evanston, Illinois. Involved in real estate appraisal and commercial lending for more than 15 years, Mr. Lucas has performed appraisals for many types of income and special purpose properties. He was previously vice president for real estate of Valuation Counselors Group, Chicago. Mr. Lucas earned a bachelor of Architecture and a master's degree in City and Regional Planning from the Illinois Institute of Technology. He is a Certified General Real Estate Appraiser and a candidate for MAI designation by the Appraisal Institute.

James C. Morell is president of Morell & Associates, Northbrook, Illinois, a healthcare consulting firm providing strategic, operational, financial, and facilities planning. With over 20 years' experience in healthcare, Mr. Morell has provided consulting services to more than 200 organizations. He was previously a partner of Ernst & Young in national practice. Mr. Morell holds a bachelor's degree in Economics and Psychology from Case Western Reserve University and a master's degree in Hospital and Health Services Administration from The Ohio State University. He is a fellow of the American Association of Healthcare Consultants and a diplomate of the American College of Healthcare Executives.

Janet R. O'Toole is a principal of O'Toole, Lucas & Associates. With 19 years' experience in real estate appraisal, Ms. O'Toole has appraised most types of commercial, industrial, residential, and special-purpose properties and frequently gives expert court testimony. She was previously a manager for real estate with Valuation Counselors Group, Chicago. Ms. O'Toole graduated from Northwestern University, summa cum laude, with a bachelor's degree in Interior Architecture and Design. She is an MAI designated member of the Appraisal Institute and a Certified General Real Estate Appraiser in the states of Illinois, Wisconsin, Indiana, Michigan, Missouri, and Kentucky.

Nancy Alfred Persily is director of strategic planning, marketing, and managed care for the George Washington University Medical Center and an assistant professor in the Department of Health Care Sciences at the university's School of Medicine and Health Science. For more than 25 years, Ms. Persily has been a recognized leader in healthcare administration, public health, healthcare planning, and eldercare. She has advised academic medical centers, teaching hospitals, community hospitals, rehabilitation facilities, nursing homes, home health agencies, and physician practice plans. Ms. Persily earned a bachelor's degree from Cornell University and an MPH from Yale University. She speaks and publishes on a wide variety of healthcare topics.

James F. Sherman, CPA, is director of financial consulting services with Zelenkofske Axelrod's Health Care Services Group. A specialist in retirement housing and long-term care with more than 25 years' experience, Mr. Sherman provides financial and management expertise in developing capital project financing techniques for healthcare and senior living institutions. His areas of concentration include market and financial feasibility studies; forecasts for community and teaching hospitals, nursing homes, and senior housing projects; and management and operational reviews for various healthcare and senior housing facilities. Mr. Sherman writes and lectures frequently on healthcare financial and managerial topics.

Index

A

AARP (American Association of Retired Persons), 15
Absorption period, 116
Access, issues of, 87–90, 95–98
Accounting policies, statement of, 195
Accrued depreciation, 117–119
Acute care, 72, 132
ADA (Americans with Disabilities Act), 87–88
ADLs (limitations in activities of daily living), 4–5, 15
Administration on Aging, 3
Aggregation, patient, 84
Agreed-upon procedures report, 176
AICPA (American Institute of Certified Public Accountants), 175, 189
Allied health services, 143
Alzheimer's disease, 5–6
  care for victims of, 11, 14–15, 17
Alzheimer's Disease and Related Disorders Association, 5, 14
American Association of Homes for the Aging, 12
American Association of Retired Persons (AARP), 15
American Healthcare Association, 10
American Housing Survey, 24
American Institute of Certified Public Accountants (AICPA), 175, 189
American Institute of Real Estate Appraisers, 133
Americans with Disabilities Act (ADA), 87–88
American Telecare, 78

Analysis
  competitive environment, 183
  facility operations, 155–156
  financial. *See* Financial feasibility study
  make-or-buy, 104–105, 110
  market. *See* Market, analysis
  post-"live", 164
  sensitivity, 177–178, 187, 196–197
Appraisal Foundation, 135
Appraisal Institute, 133, 138
Appraisal of care facilities, 115, 140
  assignment and inspection process, 128, 130–131
  date of value of property, 115–116
  design features considered in, 131–132
  qualifications of appraiser for, 137–139
  reporting methods of, 133, 136–137
  standards required in, 133–134
  and state regulations, 139–140
  types of, 135–136
  valuation methods used in, 116–117
    cost approach, 116–120
    income approach, 116, 125–130
    sales comparison approach, 120–126
Appraisers, licensure of, 138
"Arm's-length" transaction, 122
ASA (American Society of Appraisers), 138
"As-is" market value, 115
Assembly report, 176
Assets, intangible, 120
Assisted living, 15–16, 38, 132

211

Assumptions of feasibility, 173, 175, 178, 188, 195–196
Automation, limits of, 169. See also Information systems

B

Baldwin, Leo, 25
Bankruptcy remote entities, 58
Baseline financial forecast, 187
Baseline financial model, 177
Best effort transactions, 45
Beverly Enterprises, 9
Bloom, Shawn, 12
Bonds
  FSLIC insured, 52
  put or option, 53
  qualified 501(c) (3), 44, 57
  taxable, 50
  tax-exempt, 43–44
Brody, Stanley J., 9–10
Buddy System information system, 78
"Builder's method" of valuation, 119
Byers, Roy W., 205

C

Capacity management strategy, 83, 93, 107
Capital
  foreign, 51
  venture, 46, 48–49, 56–57
Capitalization, 125–126, 128
Capture rate, 184
Cardiac Alliance Company, 78
Care facility planning, 83–84, 113–114. See also Appraisal of care facilities; Financial feasibility study
  alternative planning, 109–110
  critical mass considerations in, 108–109
  decision criteria of, 102–104
  flexibility in, 91–92, 112–113
  labor resources considerations, 91, 203
  make vs. buy decision, 104–105
  in managed care marketplace, 105–107
  market determination in, 92
    integrated network, 93–94
    product-line regulatory mandates, 92–93
    sheltered services concept, 93, 103
  and market expectations
    patient and family, 94–96
    payer and employer, 98–99
    of providers, 96–98
  physical location considerations in, 87
    off-campus development, 90–92
      case study of, 101
    on-campus development
      existing facilities, 87–89
      new facilities, 89–90
    and residential support regulations, 87
  in shifting marketplace, 108
  success indicators in, 107–108
  using clinical model, 84–85, 94
    case studies of, 99–101
  and wellness continuum, 85–86
Caregivers
  expectations of, 96–98
  facilities chosen by, 14
  home care, 66–68
  long-term facility, 146–147
  strain on, 3
Care plans, individualized, 149. See also Long-term care, continuum of
Caretel Corporation, 1
CARF, 60
Case mix data, 148
Cash and investments ratio, 194
CBRFs (community-based residential facilities), 139
CCRCs (community care retirement communities), 16–17, 42–43
CDs (certificate of deposits), 52
Centralized information system, 157
Certificate-of-need (CON), 60–61, 92, 139, 148

Certificates of deposits (CDs), 52
Chronic care, 72, 85–86, 132
Clinical services continuum, 85–86, 94. *See also* Long-term care, continuum of
Cognitive impairment, 5, 15. *See also* Alzheimer's disease
Combined revenue/cost center concept, 108
Community-based residential facilities (CBRFs), 139
Community care retirement communities (CCRCs), 16–17, 42–43
Competitive environment, analysis of, 183
Compilation report, 176
CompuMed dispenser, 79
Computer-aided medical devices, 72, 77–80. See also Information systems
Concepts
  combined revenue/cost center, 108
  cost-center operating, 108
  life care, 41
  sheltered services, 93, 103
CON (certificate-of-need), 60–61, 92, 139, 148
Conditions of sale, 122
Condominium ownership, 41–42
Confidentiality, patient, 157
Consolidation, healthcare industry, 144–145
Construction loans, 51, 60
Consultants
  appraisal. *See* Appraisal of care facilities
  feasibility. *See* Financial feasibility study
Contracts
  entrance agreement, 185
  guaranteed investment, 53
Conventional lending, 50–51
Co-operatives, 41–42
Cost-center operating concept, 108
Cost estimate, 117

Cost-plus environment, 102
Countermigration, 8
Crampton, Kathleen R., 205–206
Credit enhancements
  corporate surety/guarantee, 52
  FHA insurance, 46–47, 51
  FSLIC insured bonds, 52
  guaranteed investment contracts (GICs), 53
  HUD insurance, 51–52
  letters of credit, 47–48
  private mortgage insurance, 52
  put or option bonds, 53
Critical mass considerations, 108–109
Current market value, 116

D

D' Angelo, Horace, Jr., 14
Data request list, 181–182
DCF (discounted cash flow), 126, 128
Debt financing, 58–59. *See also* Long-term care, investing in
  conventional lending, 50–51
  foreign capital, 51
  qualified 501(c)(3) bonds, 44, 57
  real estate investment trusts (REITs), 46
  securitization of mortgages, 44–45
  seller financing, 45–46
  taxable bonds, 50
  tax-exempt bonds, 43–44
Debt securitization, 58
Debt service coverage ratio, 194
Decentralized information system, 157
Delivery-Wide Health Information Network, 153
Dementia, 5. *See also* Alzheimer's disease
Demographic data, 182. *See also* Elderly population, demographics of
Departure provision, 135–136
Dependent off-campus development, 91
Depreciation, 117–119, 123

Determination-of-need (DON), 60–61, 92
Diagnosis-related groups (DRGs), 18–19, 66
Direct capitalization, 125–126
Disclosure of funds, 195
Discounted cash flow (DCF), 126, 128
Distribution management strategy, 83, 93, 107
DON (determination-of-need), 60–61, 92
Donovan, Christopher J., 206
Donovan, Michelle Regan, 202
DRGs (diagnosis-related groups), 18–19, 66
Drucker, Peter, 2
Drug infusion services, 66–68, 79
Due diligence procedures, 198

E

Economic feasibility, 158. *See also* Financial feasibility study
Economic obsolescence depreciation, 118
Economic property characteristics, comparison of, 123
Eldercare
  facilities for. *See* Long-term care, facilities for
  financing for
    government, 17–20
    private, 21–25
  long-term, 4. *See also* Long-term care
  social resources of, 3–4
  trends for future, 3. *See also* Healthcare industry
Eldercare: Positioning Your Hospital for the Future (Persily), 202
Elderly population
  demographics of, 2–3, 32–34, 70–71
  diseases of, 5
  economic status of, 6–7
  ethnic diversity of, 7–8
  geographic distribution of, 8–9, 13, 22
  health status of, 4–6
  needs of, 1–2, 9 72, 146. *See also* Elder market
  poverty in, 3, 7
  projected growth of, 6–7
  spending trends of, 6
Elder market. *See also* Elderly population
  home care, 69–71
  marketing to, 16–17, 25–26
  researching, 15–16
  trends for, 14, 26–27
Electronic medical devices, 72, 77–80. *See also* Information systems
Entrance agreement contract, 185
Environment of care facility
  caregiver expectations for, 96–97
  patient expectations for, 94–95
  payer expectations for, 98–99
Equity financing
  developer financing, 48
  equity arrangements, 41–43
  investor equity, 49–50
  limited partnership, 40–41
  participating mortgage funds, 48
  sponsored joint ventures, 41
  syndication, 49
  venture capital financing, 48–49
Evaluation of appraiser, 133
Examination report, 175

F

Feasibility
  financial. *See* Financial feasibility study
  information system, 158, 161
  project, 116
Federal Financial Institutions Reform, Recovery and Enforcement Act (FIRREA), 135
Federal Housing Administration (FHA), 38, 46–47, 51
Federal Savings and Loan Insurance Corporation (FSLIC), 52
Fee-single rights, comparison of, 121–122

FF&E (furniture, fixtures and
    equipment depreciation), 119, 123
FHA (Federal Housing
    Administration), 38, 46–47, 51
*The Financially Competitive Healthcare
    Organization* (Kaufman and Hall),
    202
Financial feasibility study, 199
    consultant, role of, 177–178, 196
    data request list for, 181–182
    levels of reporting, 175–178
    opinion letter, formats for, 190–193
    process of, 179
        consultant selection, 179–180
        examination procedures, 188–189
        financial analysis, 186–187
        information gathering, 180
        market analysis, 180, 182–186
        report preparation, 189, 194–195
        review of documents, 196–198
    purpose of, 173–174
    types of, 174
Financial statements, 189, 194
Financing. *See also* Credit
    enhancements
    debt. *See* Debt financing
    for elder care
        government, 17–20
        private, 21–25
    equity. *See* Equity financing
    permanent, 51
    tax-exempt, 43–44
    terms of, 122
FIRREA (Federal Financial
    Institutions Reform, Recovery and
    Enforcement Act), 135
501(c)(3) corporation, 44, 57
Forecasts, financial, 175–176
Foreclosure, 61
Foreign capital, 51
Form report of appraiser, 133
Franchising guidelines, 92–93
FSLIC (Federal Savings and Loan
    Insurance Corporation), 52
Functional obsolescence depreciation,
    118

Funds
    disclosure of, 195
    participating mortgage, 48
    sources of. *See* Financing
Furniture, fixtures and equipment
    depreciation (FF&E), 119, 123

G

General-use report, 176
GICs (guaranteed investment
    contracts), 53
Going-concern value, 115, 120, 125
Growth management, 83
Guaranteed investment contracts
    (GICs), 53
*Guide for Prospective Financial
    Information* (AICPA), 175, 189

H

Hall, Mark, 202
Hance, Marietta Donne, 206
HANC (home-assisted nursing
    companion), 80
HCFA (Health Care Financing
    Administration), 18, 21–24, 148
Healthcare industry
    expenditures of, 3, 70
    home care, 11–12, 65–66
        vs. hospitalization, 72–73
    impending changes in, 73–75
    market, 69–71
    opportunities in, 75–77
    payers, 68–69
    planning, 73
    prescribers, 66
    providers, 66–68
    reimbursements for, 66, 71
    technology of, 72, 77–80
    information needs of, 141–144. *See
        also* Information systems
    levels of patient care, 132
    long-term care. *See* Long-term care
    managed care programs, 71–72, 142.
        See also Medicare, managed care
        plan
    staffing issues of, 146–147, 203

Health Insurance Association of America (HIAA), 22
HealthTech Services Corporation, 80
HIAA (Health Insurance Association of America), 22
HMOs (health maintenance organizations), 21–22, 24, 62, 68–69, 71
Holbrook, Al S., 206
Home-assisted nursing companion (HANC), 80
Home care. *See* Healthcare industry, home care
Home equity conversions, 24–25
HUD (Housing and Urban Development Agency), 43, 46–47, 51–52
Hypothetical assumptions, 175

I

IADLs (limitations in instrumental activities of daily living), 4–5
IFA (Independent Fee Appraiser), 138
Income, stabilized, 125–126
Independent off-campus development, 90–91
Indicators, success, 107–108
Individual retirement accounts (IRAs), 23
Information networks, 79–80
Information systems. *See also* Computer-aided medical devices
  current industry needs, 141–142
  forces driving, 144–149
  integration, 143–144
  problems in solving, 149–151
  reimbursement, 142–143
  evaluation of
    clarifying facility goals, 152–155
    criteria checklist, 159–161
    facility operations analysis, 155–156
    systems options, 156–158, 161
  for home care, 65

implementation of, 162
  checklist for, 165–168
  managing, 168–170
  primary activities of, 163–164
Infusion therapy services, 66–68,79
Insurance
  Medicaid partnership plan, 23
  Medigap, 18–20, 22
  mortgage, 46–47, 51–52
  private, 22–23, 68
Intangible assets, 120
Integration, 61–62, 93–94, 143–145
Internal-use-only report, 176
Investing in long-term care. *See* Long-term care, investing in
IRAs (individual retirement accounts), 23
IRS (Internal Revenue Service), 43, 45, 57

J

Joint ventures, 41, 55–56
Jorgensen, Darolyn, 206–207

K

Kaufman, Kenneth, 202
Kaufman, Stephen B., 207

L

Labor resources, 91, 203
LEO, Inc., 25
Letters of credit, 47–48
Level-of-care classifications, 132–133
Licensure
  of appraisers, 138
  of care facilities, 149
  of facilities in Wisconsin, 139
  regulatory issues of, 59–61
Life care concept, 41
Lifeline Corporation, 78
Lifetime lease, 41–43
Limitations in instrumental activities of daily living (IADLs), 4–5
Limited appraisal, 135

Limited liability corporation, 57
Limited partnership, 40–41, 57
Limited-use report, 176
Limitations in activities of daily
    living (ADLs), 4–5, 15
Liquidity ratio, 194
Living developments. *See* Long-term
    care, developments for
Loan package preparation, 38–39
Location, comparison of, 122
Logan, P. Andrew, 207
Long-term care, 113–114
  continuum of, 9, 12, 73, 85–86, 94
      developments for, 31–34, 53.
    See also Care facility planning
    financial feasibility of, 173–174.
      See also Financial feasibility
      study
    funding sources for, 38–40. *See
      also* Credit enhancements;
      Debt financing; Equity
      financing
    investing in. See Long-term care,
      investing in
    motives of investors, 41, 45–46
    planning of. *See* Care facility
      planning
    product-line regulatory mandates,
      92–93
    statistics on, 36–38
    types of products, 34–35
  facilities for, 34–35, 132
    admission process of, 146
    adult day care, 11
    Alzheimer's disease units, 17. *See
      also* Alzheimer's disease
    assisted living, 15–16, 38, 132
    hospitals, 18–19, 72–73
    licensure of, 149
    management of, 146–147
    nursing homes, 12–14
    ownership of, 146
    retirement communities, 16–17
    special care units, 14–15
    staffs of, 146–147

  future of, 110–112
  and government regulations of,
    148–149
  home healthcare. *See* Healthcare
    industry, home care
  and industry consolidation, 144–145
  industry growth of, 144
  information systems for. *See*
    Information systems
  integrated networks in, 145
  investing in, 113–114. *See also* Care
    facility planning, decision
    criteria
    and choice of entity, 57–58
    cost to convert facilities, 132
    debt financing, 58–59. *See also*
      Debt financing
    joint ventures, 55–56
    legal considerations
      integration of providers, 61–62
      real estate mortgage investment
        conduits (REMICs), 62–63
      subacute care, 62
    and planning flexibility, 113
    regulatory issues of
      certificates, 60–61
      letters and reports, 60
      licenses, 59–60
    traditional, 57
    venture capital, 56–57
  managed, 145
  market changes in, 145
  operational characteristics of, 146–
    148
  scope of, 146
  services for, 9–10, 66–68
  trends in, 144–145
Long-Term Care NEWS (McKnight),
    201
LTLTC (long-term long-term care), 10
Lucas, Michael A., 207

M

MAI (Member of the Appraisal
    Institute), 138

Mainframe computers, 156–157. *See also* Information systems
Make-or-buy analysis, 104–105, 110
Managed care
  marketplace, 105–107
  programs, 71–72, 142. *See also* Medicaid; Medicare
Management strategies, 83, 93, 107, 146–147
Market. *See also* Elder market
  analysis of, 116, 177–178, 180, 182–186
  changes in long-term, 145
  conditions, comparison of, 122
  demand, assessment of, 186
  determination of, 92–94, 103
  expectations of, 94–99
  home care, 69–71
  managed care, 105–107
  niching, 109–110
  penetration rate of, 183–184
  primary area of, 180, 182
  securitization, 62
  value, 115–116
Marketing to Baby Boomers and Beyond (Wolfe), 16–17
Market-rate housing, 184
Matson, Theodore A., 202
MDSs (minimum data sets), 149, 154, 157
Means test-triggered entitlement, 20
Medicaid
  costs covered by, 10, 68
  enactment of, 1
  insurance partnership plan of, 23
  managed care plan, 71–72
  program, 20–21, 147
  provider certification, 60–61
  recipients of, 139–140
  reimbursement system of, 148, 162
  states' policies for, 13–14, 147
Medicare
  costs covered by, 10, 68
  costs not covered by, 79
  diagnosis-related reimbursements of, 18–19
  enactment of, 1
  and home care agencies, 70
  managed care plan of, 21–22, 24, 68–69, 71–72
  and physicians' fees, 19
  program, 17–18, 147
  provider certification, 60–61, 147
  reimbursement system of, 68, 71, 148
Medicare Catastrophic Care Act, 18
Medicare Select, 22
Medication reminders, 79
Medigap insurance, 18–20, 22
Member of the Appraisal Institute (MAI), 138
Membership organizations, 41–42
Mills, Ronald E., 206
Minicomputers, 156–157. *See also* Information systems
Minimum data sets (MDSs), 149, 154, 157
Models, financial, 186–187
Morell, James C., 208
Mortgages
  insurance for, 46–47, 51–52
  participating funds, 48
  reverse, 24–25
  securitization of, 44–45

N

Narrative report of appraiser, 133–134
National Association for Home Care, 11
National Investment Conference (NIC), 201
Net operating income (NOI), 126
NHPI (National Health Provider Inventory), 12–13
Niching, 109–110
NIC (National Investment Conference), 201
NOI (net operating income), 126
Nonrealty components, comparison of, 123
North American Securities Administration Association, 46

Nosocomial infections, 72–73, 97
Notes to financial report, 196
Not-for-profit corporations, 56–57. *See also* Tax-exempt financing
Nursing services, 66–68, 146

O

OBRA (Omnibus Reconsilation Act), 148–149
Off-campus development, 90–92, 101
Offerings
  private, rules for, 49–50
  public, 176, 178, 196–198
  statement of, 196–198
Omnibus Reconcilation Act (OBRA), 148–149
On-campus development, 87–90
Operating expense ratio, 123
Operational feasibility, 158
Opinion letter, feasibility study, 189–193
O'Toole, Janet R., 208
Outpatient Case Management: Strategies for a New Reality (Donovan and Matson), 202

P

Patient aggregation, 84
PC-based distributed networks, 156–157. *See also* Information systems
Penetration rates, 183–184
Permanent financing, 51
Persily, Nancy Alfred, 202, 208
Personal Companion computer aid, 79
PERS (personal emergency response systems), 77–78
Physical deterioration depreciation, 118
Physical property characteristics
  comparison of, 122–123
  inspection of, 128, 130–131
Physicians as prescribers, 66
Planning of facilities. *See* Care facility planning

Platforms, computer operating, 156–157
Policies
  state Medicaid, 13–14, 147
  statement of accounting, 195
Post-"live" analysis, 164
PPO (preferred provider organization), 22, 71
PPS (prospective payment system), 18–19, 71, 102
Preliminary market analysis, 177–178
Premarketing, 184–185
Primary market area, 180, 182
Principal transactions, 45
Private offering, rules for, 49–50
Products, long-term care, 34–35, 92–93, 109–110
Projections, financial, 175–176
Project steering committee, 155
Property rights conveyed, 121–122
Property use, 123
Proprietary system platforms, 150
Prospective financial information, 174–175
Prospective market value, 116
Prospective payment system (PPS), 18–19, 71, 102
Public offerings, 176, 178, 196–198

R

RAI (resident assessment instrument), 148–149, 151, 154
RAPs (resident assessment protocols), 149, 151, 154
Rated securities, 44, 63
Ratios, financial, 194
RBRVS (Resource-Based Relative Value Scale), 19
Real estate investment trusts (REITs), 46
Real estate mortgage investment conduits (REMICs) 62–63
Regulations. *See also* Medicaid; Medicare
  appraisal considerations of, 139–140

government, 148–149
licensing, Wisconsin, 139
planning considerations of, 87
Regulatory approval process, 178
Rehabilitation services, 67–68, 86
Reimbursement considerations
   for home care services, 66, 71
   and information systems, 142–143
   Medicaid system, 148, 162
   Medicare system, 18–19, 68, 71, 148
REITs (real estate investment trusts), 46
REMICs (real estate mortgage investment conduits), 62–63
Reports
   agreed-upon procedures, 176
   appraisal, 133–134, 136–137
   assembly, 176
   compilation, 176
   examination, 175
   financial feasibility, 175–178, 189, 194–195
   general-use, 176
   internal-use-only, 176
   limited-use, 176
   notes to financial, 196
   regulatory, 60
   restricted, 136–137
   self-contained, 136
   summary, 136
Request for information (RFI), 158
Resident assessment instrument (RAI), 148–149, 151, 154
Resident assessment protocols (RAPs), 149, 151, 154
Resident care revenue streams, 148
Resident income requirements, 43–44
Resource-Based Relative Value Scale (RBRVS), 19
Respiratory therapy, 66–68
Restricted report, 136–137
Retirement Migration Project, 8
Revenue-focused care products, 109–110
Reverse mortgages, 24–25
RFI (request for information), 158
Robert Wood Johnson Foundation, 23

S

Sale, conditions of, 122
S corporation, 57
SCUs (special care units), 14–15
SEC (Securities and Exchange Commission), 46, 49–50
Securities laws, 49–50
Securitization
   debt, 58
   market, 62
   of mortgages, 44–45
Self-contained report, 136
Senility, 5. *See also* Alzheimer's disease
Senior Living and Long Term Care Industries, 201
Senior Real Property Appraiser (SRPA), 138
Sensitivity analysis, 177–178, 187, 196–197
Sheltered services concept, 93, 103
Sherman, James F., 208–209
S/HMOs (social health maintenance organizations), 24
Short-term long-term care (STLTC), 9–10
Single-system monitoring, 78
SNF (skilled nursing facility), 17
Social health maintenance organizations (S/HMOs), 24
Social Security, 1, 6
Special care units (SCUs), 14–15
Special-use properties, valuation of, 119–120
SRPA (Senior Real Property Appraiser), 138
Stabilized operating statement, 125, 127
Standards of Professional Practice, 133
STLTC (short-term long-term care), 9–10
Strategies, management, 83, 93, 107, 146–147
Subacute care, 62, 132
Suitability review, 59–60

Summary report, 136
Syndication, 49

T

Takeout loan, 51
Tax-exempt bond offering, 198
Tax-exempt financing, 43–44
Technical feasibility, 158
"Time" adjustment, 122
Tokos, 78
Traffic control, 88–89, 97
Transactions
  "arm's-length", 122
  best effort, 45
  principal, 45
Transceptor Technologies, Inc., 79

U

Uniform Standards of Professional Appraisal Practice (USPAP), 117, 133, 135, 137–138
United States government
  Administration of Aging, 3
  Americans with Disabilities Act, 87–88
  antikickback legislation of, 56
  Department of Health and Human Services, 21
  Federal Financial Institutions Reform, Recovery and Enforcement Act, 135
  Federal Housing Administration, 38, 46–47, 51
  Federal Savings and Loan Insurance Corporation, 52
  General Accounting Office, 34
  Health Care Financing Administration, 18, 21–24, 148
  Housing and Urban Development Agency, 43, 46–47, 51–52
  Housing and Urban-Rural Recovery Act, 46–47
  Internal Revenue Service, 43, 45, 57
  Joint Commission on Accreditation of Health Care Organizations, 60

Medicaid. *See* Medicaid.
Medicare. *See* Medicare
Medicare Catastrophic Care Act, 18
Medigap insurance of, 18-20, 22
Omnibus Reconciliation Act, 148–149
prospective payment system of, 18–19, 71, 102
Rehabilitation Act, 87–88
Securities Act, 50
Securities and Exchange Commission, 46, 49–50
Tax Reform Act, 49
University of Minnesota, study by, 15
Use-rate formulas, 93
USPAP (Uniform Standards of Professional Appraisal Practice), 117, 133, 135, 137–138
Utilization guidelines, 149

V

Valuation, 119–120. *See also* Appraisal of care facilities, valuation methods used
Venture capital, 46, 48–49, 56–57
Viability, financial. *See* Financial feasibility study
Video, medical uses of, 78
Vital sign monitoring devices, 78
VNA (Visiting Nurse Association), 66
Voice-activated computer aids, 79

W

Wellness continuum, 85–86, 94. *See also* Long-term care, continuum of
Wheelchair access, 87–88
Wisconsin, licensing regulations of, 139
Wolfe, David, 16–17

Y

Yield capitalization, 125–126